SYSTEM SPECIFICATION
& DESIGN LANGUAGES

T0205347

System Specification & Design Languages

Best of FDL'02

Edited by

Eugenio Villar
University of Cantabria, Spain

and

Jean Mermet
KeesDA, France

KLUWER ACADEMIC PUBLISHERS
BOSTON / DORDRECHT / LONDON

A C.I.P. Catalogue record for this book is available from the Library of Congress.

ISBN 978-1-4419-5348-3
e-ISBN 978-0-306-48734-7

Published by Kluwer Academic Publishers,
P.O. Box 17, 3300 AA Dordrecht, The Netherlands.

Sold and distributed in North, Central and South America
by Kluwer Academic Publishers,
101 Philip Drive, Norwell, MA 02061, U.S.A.

In all other countries, sold and distributed
by Kluwer Academic Publishers,
P.O. Box 322, 3300 AH Dordrecht, The Netherlands.

Printed on acid-free paper

All Rights Reserved
© 2003 Kluwer Academic Publishers, Boston
Softcover reprint of the hardcover 1st edition 2003
No part of this work may be reproduced, stored in a retrieval system, or transmitted
in any form or by any means, electronic, mechanical, photocopying, microfilming, recording
or otherwise, without written permission from the Publisher, with the exception
of any material supplied specifically for the purpose of being entered
and executed on a computer system, for exclusive use by the purchaser of the work.

Contents

Contributing Authors

Luciano Baresi, Jens Bastian, A.J.W.M. ten Berg, Morgan Björkander, Massimo Bombana, Francesco Bruschi, Jerry R. Burch, Lukai Cai, Francky Catthoor, Rong Chen, Ernst Christen, Christoph Clauss, Wim Dehaene, Geert Deconinck, Florent de Dinechin, Karsten Einwich, Hilding Elmqvist, Antonio Acosta Fernández, William Fornaciari, Masahiro Fujita, Daniel Gajski, Christoph Grimm, Joachim Haase, Stefan Hallerstede, Stefaan Himpe, Sorin A. Huss, Gjalt de Jong, Stephan Klaus, Cris Kobryn, Paul Kritzinger, Oded Lachish, Luciano Lavagno, Christophe Lallement, Natividad Martínez Madrid, M. Manjunathaiah, Roel Marichal, Grant Martin, Sven Erik Mattsson, Jef van Meerbergen, Trevor Moore, Dieter Monjau, Felipe Ruiz Moreno, Elisabetta di Nitto, Mike Olivarez, Paul van Oostende, Martin Otter, Roberto Passerone, Marc Pauwels, François Pêcheux, Luigi Pomante, P.H.A. van der Putten, Jan Rabaey, Sven Reitz, Tanguy Risset, Ralf Rosenberger, Thanyapat Sakunkonchak, Mohamed Abdel Salam, Ashraf Salem, Alberto Sangiovanni-Vincentelli, Josephus van Sas, Peter Schwarz, Donatella Sciuto, Ralf Seepold, Marco Sgroi, Geert Sonck, Mike Spivey, Mathias Sporer, B.D. Theelen, Timo Trautmann, Yves Vanderperren, J.P.M. Voeten, Klaus Waldschmidt, F.N. van Wijk, Avi Ziv

Preface

The Forum on Design Languages (FDL) is the European Forum to exchange experiences and learn of new trends, in the application of languages and the associated design methods and tools, to design complex electronic systems. By offering several co-located workshops, this multi-faceted event gives an excellent opportunity to gain up-to-date knowledge across main aspects of such a wide field. All the workshops address as their common denominator the different application domains of system-design languages with the presentation of the latest research results and design experiences.

FDL'02 was organized as four focused workshops, Languages for Analog and Mixed-Signal system design, UML-based system specification and design, C/C++-based system design and Specification Formalisms for Proven design.

FDL served once more as the European Forum for electronic system design languages and consolidates as the main place in Europe where designers interested in design languages and their applications can meet and interchange experiences.

In this fourth book in the CHDL Series, a selection of the best papers presented in FDL'02 is published. The book contains outstanding research contributions in the four areas mentioned above. So, The Analog and Mixed-Signal system design contributions cover the new methodological approaches like AMS behavioral specification, mixed-signal modeling and simulation, AMS reuse and MEMs design using the new modeling languages such as VHDL-AMS, Verilog-AMS, Modelica and analog-mixed signal extensions to SystemC.

UML is the de-facto standard for SW development covering the early development stages of requirement analysis and system specification. The UML-based system specification and design contributions address latest results on hot-topic areas such as system profiling, performance analysis and UML application to complex, HW/SW embedded systems and SoC design.

C/C++-for HW/SW systems design is entering standard industrial design flows. Selected papers cover system modeling, system verification and SW generation.

The papers from the Specification Formalisms for Proven design workshop present formal methods for system modeling and design, semantic integrity and formal languages such as ALPHA, HANDLE and B.

The book is of interest to researchers in system specification and design languages and associated methods and tools, system specification and design engineers and post-Graduate lecturers and students involved in advanced courses on system specification and design.

Eugenio Villar and Jean Mermet

Previous books

Jean Mermet (Editor), "Electronic Chips & Systems Design Languages", Kluwer, 2001.
Jean Mermet, Peter Ashenden and Ralf Seepold (Editors), "System-on-Chip Methodologies and Design Languages", Kluwer, 2001.
Anne Mignotte, Eugenio Villar and Lynn Horobin (Editors), "System-on-Chip Design Languages", Kluwer, 2002.

1

PART I: ANALOG AND MIXED-SIGNAL SYSTEM DESIGN

Chapter 1

SELECTED TOPICS IN
MIXED-SIGNAL SIMULATION
Invited paper

Ernst Christen
Synopsys, Inc., Hillsboro, OR 97124
Ernst.Christen@synopsys.com

Abstract: We give an overview of mixed-signal simulation. We discuss properties and limitations of algorithms and provide suggestions, useful for implementers and model writers, for robust mixed-signal simulation.

Key words: Mixed-signal simulation, simulation algorithms, differential-algebraic equations.

1.   INTRODUCTION

With the increasing popularity of mixed-signal hardware description languages, mixed-signal simulation has become a reality for many designers. Little has been written about what mixed-signal simulation is and how it works, but this information is important for designers to understand their simulation results.

In this contribution, we give an overview of mixed-signal simulation, the simulation of a model that contains a continuous-time portion and an event-driven portion. Each portion is simulated using its specialized algorithm, and the two portions interact. We discuss algorithms for the two portions and their interaction. We describe strengths and limitations of these algorithms and make several suggestions to address the limitations as well as potential simulation problems. We use VHDL-AMS terminology and definitions from its Language Reference Manual (LRM) [1], but the discussion applies to mixed-signal simulation driven by any language.

E. Villar and J. Mermet (eds.), System Specification and Design Languages, 5–17.
© 2003 *Kluwer Academic Publishers.*

In section 2 we discuss differential-algebraic equations: their properties, their formulation, and algorithms for their numerical solution. Section 3 describes the differences of an event-driven simulation algorithm used for mixed-signal simulation compared to the same algorithm used for pure digital simulation. Section 4 presents several aspects of the interaction of the two classes of algorithms, including the VHDL-AMS simulation cycle, synchronization methods, discontinuity handling, and initialization aspects.

2.   DIFFERENTIAL-ALGEBRAIC EQUATIONS

The continuous-time portion of a mixed-signal system is described by differential-algebraic equations (DAEs) of the form:

$$F(t, y, \dot{y}) = 0 \tag{1}$$

where t denotes time, y denotes the unknowns, and $\dot{y}$ denotes the derivative of the unknowns with respect to time. In general, the system of equations $F = 0$ consists of two parts: a set of ordinary differential equations (ODEs), and a set of algebraic equations. If there are no algebraic equations, eqn. (1) describes a system of ODEs, that is, DAEs include ODEs as a special case.

Given a set of initial conditions

$$y(t_0) = y_0 \tag{2}$$

it is the task of the simulator to find values for the unknowns y and their derivatives w.r.t. time such that eqn. (1) holds for all $t \geq t_0$.

In the remainder of this section we will first discuss how the DAEs are constructed and then give an overview of the numerical methods for the solution of DAEs. We will assume that the solution is unique.

2.1   Formulation of Differential-Algebraic Equations

DAEs arise naturally when describing the behavior of continuous systems. The DAEs may be defined directly if the system can be described as a whole. However, this approach is limited to small systems due to the complexities involved.

A more flexible approach is to describe the system as a hierarchy of component instances, each containing the description of a small portion of the system. This approach is of particular interest in the context of designs described by a mixed-signal hardware description language, where it is the task of the elaborator to construct the DAEs from statements in the language describing each component, the interconnection of the component instances, and the language semantics.

In the VHDL-AMS language the continuous-time behavior of a model is described using the class of simultaneous statements. The simple simultaneous statement, whose form is

$$lhs\_simple\_expression == rhs\_simple\_expression; \qquad (3)$$

allows a model writer to express a relationship between quantities. Each such statement defines a characteristic expression

$$rhs\_simple\_expression_i - lhs\_simple\_expression_i \qquad (4)$$

for each scalar subelement of the simple expression. It is the task of the simulator to find values for the quantities such that the value of each characteristic expression is close to zero, thereby enforcing the desired relationship between the involved quantities. The simultaneous procedural statement provides a way to write characteristic expressions using a procedural style. The simultaneous if statement and the simultaneous case statement provide means to select a collection of characteristic expressions based on the value of a condition. As a result, the characteristic expressions in effect may change over the duration of a simulation.

The VHDL-AMS LRM [1] calls the characteristic expressions defined by simultaneous statements *explicit*. It also defines one implicit characteristic expression for each scalar subelement of a port association whose formal is a quantity or a terminal; for terminal ports these implicit characteristic expressions describe the conservation laws (KCL and KVL in electrical systems). Further, the LRM defines one characteristic expression for each scalar subelement of each branch across quantity declaration. Finally, source quantities and the quantity attributes Q'Dot, Q'Integ, and Q'Delayed each imply a characteristic expression for every one of its scalar subelements; these characteristic expressions are defined by the augmentation sets specified in the LRM.

The reader will notice that there is a direct correspondence between eqn. (1) and the characteristic expressions implied by the simultaneous statements and declarations in the design units bound to the component instances in a system and the interconnections between these component instances. The language definition guarantees that the number of characteristic expressions equals the number of quantities, which, in turn, equals the number of unknowns; this number is quite large. In general, the DAEs defined by eqn. (1) are very sparse because each characteristic expression involves only a small number of the unknowns. Methods are available to reduce the size of eqn. (1) by applying suitable topological transforms (see, for example, [7]), but their description is beyond the scope of this contribution.

There is no requirement that the expressions in eqn. (3) be continuous over time nor that different sets of characteristic expressions selected for different values of a condition produce a solution that is smooth at the point where the selection changes. We can, however, state that in the most general case the DAEs in eqn. (1) are controlled by a finite state machine (implemented by the simultaneous if and case statements) [8] and that the solution of eqn. (1) is piecewise continuous over time.

2.2   Numerical Solution of Differential-Algebraic Equations

The traditional approach to solving DAEs has been to first transform the DAEs to ODEs by eliminating the algebraic equations and then to apply one of the ODE solvers. Several simulation programs used by the control system community are still based on this approach. Unfortunately, reducing DAEs to ODEs may produce less meaningful variables in the transformed equations, and it often destroys the sparsity of the problem [4].

To overcome these problems, numerical mathematicians began to study the properties of DAEs and methods for their direct solution in the early 1970s. The results have been presented in a monograph [4], although newer results are available. We will summarize these results in the remainder of this section and draw some conclusions for practical implementations.

2.2.1   Classification of Differential-Algebraic Equations

The most important theoretical concept in classifying DAEs is their *index*. The index of a DAE is defined as the minimum number of times that any part of eqn. (1) must be differentiated w.r.t. t in order to determine $\dot{y}$ as a function of y [4]. This process would produce the ODE

$$\dot{y} = G(t, y) \tag{5}$$

where G is a function of the derivatives of F. The solution of eqn. (1) is also a solution of eqn. (5), but the latter has additional solutions. We observe that, unlike the solution of ODEs, the solution of DAEs may involve derivatives of input functions (stimulus) and coefficients.

The importance of the index becomes apparent when considering the theoretical result that DAEs with index greater than 2 cannot reliably be solved with variable step size DAE solvers and that DAEs with index 2 can only be solved if they are in the so-called semi-explicit form

$$E\ddot{y} = H(t, y) \tag{6}$$

where E is a matrix that for DAEs is singular. To solve higher index DAEs the recommendation is to transform the DAEs to an index 1 system.

2.2.2 Algorithms for the Numerical Solution of Differential-Algebraic Equations

All numerical solution methods compute solutions of the DAEs at discrete time points t_i, the *analog solution points* (ASPs). Moreover, all methods are based on the assumption that one or more solutions of the DAE exist. We denote the times of the ASPs with $t_{n-k}, t_{n-k+1}, ..., t_{n-1}$, and the corresponding solutions with $y_{n-k}, y_{n-k+1}, ..., y_{n-1}$, and we are interested in finding the ASP y_n at time t_n.

Two classes of algorithms have been studied extensively for the numerical solution of DAEs [4]: the linear multi-step (LMS) methods, and the one-step methods. A k-step LMS method uses information from k previous ASPs to compute the ASP at t_n, while one-step methods require the existence of only one past ASP. Most implementations use variations of these methods, but there are some implementations that combine LMS with one-step methods to take advantage of their combined strengths.

Linear Multi-Step Methods. LMS methods are by far the most frequently used methods for solving DAEs. For fixed time steps $h = t_i - t_{i-1}$, a k-step LMS method can be described by [3]

$$y_n = \sum_{i=0}^{k-1} \alpha_{k-i} y_{n-k+i} + h \sum_{i=0}^{k} \beta_{k-i} \dot{y}_{n-k+i} \tag{7}$$

i.e. the solution is approximated by a polynomial of order p. The coefficients α_i and β_i are selected based on criteria like stability of the method, use of the method as predictor or corrector formula (see below), approximation error, and others. Well known examples of LMS methods are:

Method	Formula	k	p
Forward Euler	$y_n = y_{n-1} + h\dot{y}_{n-1}$	1	1
Backward Euler	$y_n = y_{n-1} + h\dot{y}_n$	1	1

Method	Formula	k	p
Trapezoidal	$y_h = y_{n-1} + \dfrac{h}{2}(\dot{y}_{n-1} + \dot{y}_n)$	1	2
Second order Gear	$y_n = \dfrac{4}{3}y_{n-1} - \dfrac{1}{3}y_{n-2} + \dfrac{2h}{3}\dot{y}_n$	2	2

Note that forward and backward Euler are traditionally considered LMS methods although they are also one-step methods. Among the LMS methods, the backward differentiation formulae (BDF), characterized by $\beta_i = 0$ $\forall i > 0$, are the most popular. BDFs are stable for $k \leq 7$. They are used most often with a variable step size. In general, higher order methods can take larger time steps than lower order ones for the same error criteria. Therefore, most solvers also vary the order of the LMS method: they start with a low order method at the beginning of a continuous interval and increase the order for subsequent ASPs. The choice of the method for a particular order depends on the problem to be solved and is typically left to the user.

As indicated by eqn. (7), LMS methods only need information from past ASPs to determine the ASP at t_n. This distinguishes them from one-step methods, discussed next, which need function values and derivatives at times between t_{n-1} and t_n in addition to the information from the ASP at t_{n-1}. This distinction becomes important when the function evaluation (i.e., the evaluation of the characteristic expressions) is expensive. It is, besides the relatively simple implementation, the main advantage of the LMS methods.

One-Step Methods. Besides the one-step methods already discussed, the implicit Runge-Kutta (IRK) methods are the most studied one-step methods. With $h = t_n - t_{n-1}$, an M-stage IRK method applied to eqn. (1) is given by [4]

$$F(t_{n-1} + c_i h, Y_i, \dot{Y}_i) = 0, \; i = 1, 2, \cdots M$$

$$Y_i = y_{n-1} + h \sum_{j=1}^{M} a_{ij} \dot{Y}_j \qquad\qquad (8)$$

$$y_n = y_{n-1} + h \sum_{i=1}^{M} b_i \dot{Y}_i$$

For $M = 1$ and suitable values for a, b, c, eqn. (8) describes a backward Euler method. IRK methods have the advantage that the integration order after a discontinuity. Conversely, LMS methods must be restarted at low order after a discontinuity because no past values except y_{n-1} are available.

Error Handling and Time Step Control. With the exception of forward Euler, all methods described above are implicit methods. That is, in order to determine y_n they also need $\dot{y}_n$, which can be obtained from eqn. (7) or eqn. (8). Substituting $\dot{y}_n$ into eqn. (1) yields a system of nonlinear equations in y_n that can be solved using, for example, Newton's method. A suitable start value y_n^P for y_n can be determined using a *predictor* method, which typically is an LMS method with $\beta_0 = 0$. The implicit methods described above are then called *corrector* methods; their result is the corrector value y_n^C.

The traditional method of estimating local errors in the solution of ODEs is based on the difference between the corrector and the predictor values:

$$\varepsilon = \left\| c(y_n^C - y_n^P) \right\| \tag{9}$$

where c depends on the method used. For the numerical solution of DAEs this criterion is often modified to only include the unknowns y_i whose corresponding derivative w.r.t. time, $\dot{y}_i$, is part of eqn. (1). The error estimate is used for two purposes. If ε is small enough, the solution y_n is accepted as an ASP, and ε is used to control the size of the next time step. Otherwise, the solution y_n at time t_n is rejected and a new solution is determined at $t_n' < t_n$. This process may be repeated if the error is still too large. Unfortunately, after repeated reduction of the step size, the corrector may start to diverge because the condition of the Jacobian becomes increasingly poor. This is best demonstrated by solving a linear semi-explicit DAE $E\dot{y} = Hy$ using backward Euler. With $\dot{y}_n = (y_n - y_{n-1})/h$ we obtain

$$\left(\frac{E}{h} - H\right) y_n = \frac{E}{h} y_{n-1} \tag{10}$$

With decreasing h, E becomes more and more dominant in the matrix $(E/h - H)$, but E is singular! We conclude that repeated reduction of the step size may not be a good strategy unless the condition number of the matrix is considered in the algorithm.

A different problem may occur in the context of an unexpected discontinuity. Petzold showed, using an index 2 DAE as an example, that the error estimate may become independent of the step size if there is no ASP at the time of the discontinuity [5]. This means that, contrary to expectations, a reduction of the step size does not reduce the error. The problem is resolved by computing an ASP at the time of the discontinuity. This imposes the requirement on the model that causes the discontinuity to communicate this fact to the solution algorithm. In VHDL-AMS this is accomplished by means of the **break** statement.

The above considerations are based on local error estimates. Gear et.al. showed [9] that under mild conditions the global error in the numerical solution of a k-step BDF with variable step size applied to a constant coefficient DAE with index v is $O(h_{max}^q)$, where $q = \min(k, k - v + 2)$. This means that the solution of an index 3 DAE with backward Euler will have $O(1)$ error! This is important because analog solvers based on BDF start the integration with a first order method. It indicates that an index 3 DAE cannot be started, or restarted after a discontinuity, with a BDF.

2.2.3   Practical Considerations

Given the importance of the index of a DAE and its relation to the solvability of the DAE, the question arises what an implementation or a model writer can do to get reliable simulation results.

The best advice to a model writer is to choose the right coordinate system to describe a problem. For example, the equations describing the motion of a pendulum, expressed using Cartesian coordinates have index 3, but the same problem expressed using polar coordinates is a simple ODE (i.e., index 0) [4]. Unfortunately, this advice may be difficult to follow, for two reasons: First, for a system composed of many different model instances, it may be very difficult to understand the implications of a modeling decision in a model on the index of the problem to simulate. Second, the index of some problems is inherently high. For example, the equations describing linear motion of a mass have index 0 if the acceleration of the mass is defined, index 1 if its velocity is defined, and index 2 if the position is defined.

For implementations the obvious thing to do is to be aware of DAEs and adjust the algorithms accordingly. We would expect the next step to be a determination of the index of the DAEs to solve. However, this is not very practical because an index determination is relatively expensive [4]. In addition, the index may change during a simulation due to the conditional statements in the language (for VHDL-AMS: simultaneous if and case statements). For the same reasons, index reduction is also not very practical; additionally, the solution of the reduced index system may drift [4].

3.   EVENT-DRIVEN TECHNIQUES

Algorithms for the simulation of event-driven systems are well understood. To support mixed-signal simulation, few changes are necessary. The most important change affects the representation of time. In event-driven simulators time is typically represented as an integer multiple of some time unit, the resolution limit (in VHDL: 1 fs by default [2]). This discrete representation provides time values with a constant absolute precision. For continuous-time algorithms a floating-point representation of time with a constant relative precision is more suitable. In a mixed-signal simulation, threshold events (see below) occur frequently between two discrete times. To maintain the exact timing of such events, an implementation must therefore adapt its event handling accordingly. In VHDL-AMS a similar requirement exists to support **wait** statements with a timeout clause of type Real. All other events, including events scheduled at a non-discrete time and with non-zero delay, will again occur at discrete times.

4. MIXED-SIGNAL SIMULATION

We now focus on uniting continuous-time and event-driven algorithms for a mixed-signal simulation. We first revisit the VHDL-AMS simulation cycle as an example of a mixed-signal simulation cycle. Next, we discuss methods to synchronize the continuous-time algorithm and the event-driven algorithm. We then describe the handling of discontinuities, and finally we look at the initialization of a mixed-signal simulation.

4.1 The VHDL-AMS Simulation Cycle

The VHDL-AMS simulation cycle specifies how time advances during a simulation. Ignoring portions related to postponed processes, discontinuity handling and the Domain signal, it can be summarized as follows:

a) The analog solver determines a sequence of ASPs at times T_i in the interval $[T_c, T_n']$, where T_n' is the smaller of T_n and the earliest time in the interval $[T_c, T_n]$ at which the sign of the expression $Q - E$ for any signal Q'Above(E) changes. The times T_i must include T_n'.

b) The current time T_c is set to T_n'.

c) Each active signal in the model is updated.

d) Each process that is sensitive to a signal that had an event in this simulation cycle resumes.

e) Each process that has resumed in the current simulation cycle is executed until it suspends.

f) T_n is set to the earliest of:
   - The value of type universal_time corresponding to Time'High,
   - The next time at which a driver becomes active, or
   - The next time at which a process resumes.

   If $T_n = T_c$, then the next simulation cycle will be a *delta cycle*.

4.2   Interaction of Continuous-Time and Event-Driven Algorithms

When the analog solver resumes at time T_c, we know that the values of the signals will not change before some time T_n', which initially is set to T_n. This means that even if the DAEs depend on signal values, they will be stable in this respect up to T_n'. We know, however, that at time T_n' either a signal will be updated or a process will resume, which may cause a signal update at that time, delayed by one delta cycle. The signal may be an "ordinary" signal (T_n' is left unchanged), or it may be a signal of the form Q'Above(E), in which case T_n' is reset to the time at which the sign of the expression $Q - E$ changes (colloquially, when Q crosses the threshold E).

The analog solver relinquishes control when it has determined the ASP at T_n'. The kernel process first updates the current time T_c to T_n'. Next, it updates the active signals, and then the processes that are sensitive to signals that had an event in the current simulation cycle resume. When they execute in the next step of the simulation cycle, the value of any quantity read in a process is correct because the time T_n' of the last ASP is the time at which the process executes. The cycle repeats after a new T_n has been determined.

Threshold Detection. When the analog solver has determined an ASP at time T_i, it must detect whether any thresholds were crossed in the interval $[T_{i-1}, T_i]$. A necessary (for first order methods only) and sufficient condition for a threshold crossing is that the expression $Q - E$ has different signs at T_{i-1} and T_i. Once the presence of a threshold crossing has been established, a root finding method can be used to find the time of the crossing [10]. The LRM allows for a small error in the time of the crossing, but requires that the time of the threshold event be greater than or equal to the exact crossing time.

T_n' is set to the earliest time any threshold is crossed between T_{i-1} and T_i. Then the ASP at T_i, which was only tentative, is rejected and a new ASP at T_n' is determined. A patented optimization [6], part of the Calaveras® algorithm, computes the ASP at T_n' by interpolation using an interpolation polynomial of the same order as the LMS method that yielded the ASP at T_i. The properties of the LMS methods guarantee sufficiently accurate results using this approach at a considerably lower cost than a solution of the DAEs.

4.3   Discontinuities

Assume that the analog solver has determined the ASP at T_n' using an *m*-th order LMS method. When it resumes in the next simulation cycle, it must be able, for efficiency reasons, to use its past history and continue with the *m*-th order method or even increase the order. However, we also know that if a signal update causes a discontinuity, we will have to restart the algorithm with a low order method. How does the analog solver know when to restart?

While it is conceptually possible for the analog solver to determine whether an event occurred on a signal on which the DAEs depend, the VHDL-AMS language requires that a model notify the analog solver of a need to restart the algorithm. This choice was made because an approach based on events alone would be unreliable: on the one hand, discontinuities unrelated to events on signals may occur, on the other hand, an event on a signal may not cause a discontinuity even if the DAEs depend on the signal.

The VHDL-AMS definitions related to discontinuity handling are based on energy conservation. By default, a quantity Q keeps its value across a discontinuity if its derivative w.r.t. time, Q'Dot, appears in the model. This has the effect that physical quantities with storage effects, such as charge and flux, don't change discontinuously unless requested by an initial condition. The definitions can be used to construct a system of equations to determine a consistent set of values for the beginning of each continuous interval.

An algorithmic approach to handling discontinuities is based on a proof by Sincovec et.al. that a k-step constant step size BDF applied to a constant coefficient DAE of index v is convergent of order $O(h^k)$ after $(v-1)k+1$ steps [11]. This result was used by Sincovec and later by Vlach et.al. [12] to determine the solution of a DAE at the beginning of a continuous interval by taking several steps into the interval and then stepping backward to the beginning of the interval (i.e. to the discontinuity) using the same steps. This is illustrated for an index 2 example in the following figure.

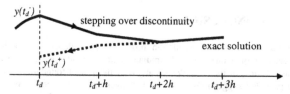

The solid line depicts integration with positive step size, the dashed line, integration with negative time steps. $y(t_d^-)$ is the value of the unknown $y(t)$ immediately before the discontinuity. According to the above result, the backward Euler method converges to the exact solution after two steps, i.e. at t_d+2h. Integrating backward using the same steps yields correct solutions, including $y(t_d^+)$, the initial value of $y(t)$ after the discontinuity.

4.4   Initialization

The simulation cycle requires that each signal and quantity in the model have a well-defined value. Such values are established during initialization, which for VHDL-AMS can be summarized as:
1. Compute the driving value and effective value of each explicit signal.
2. Set each implicit signal and each implicit quantity to its initial value.
3. Execute each process in the model until it suspends.
4. Determine the DC equations, then apply initial conditions.
5. Set T_n, the time of the next simulation cycle to 0.0.

Initialization is immediately followed by the first step of the simulation cycle: the execution of the analog solver. We make two observations:
- During the first execution of a process, at step 3. of the initialization, each quantity has its initial value.
- During the first execution of the analog solver, each signal has its initial value.

The second observation can explain a potential failure of the first execution of the analog solver. The failure is related to quantities of the form S'Ramp and S'Slew, whose initial value is the value of their prefix, S, which must be of type Real. Unless defined by an initial value expression, the initial value of S is Real'Left, the largest negative floating-point number. If values this large appear in a characteristic expression, an attempt to solve the DAEs numerically will likely result in an overflow, and the solution fails. The remedy is simple: Always specify an initial value for signals that appear as the prefix of the attributes 'Ramp or 'Slew.

5.   CONCLUSIONS

Robust mixed-signal simulation poses many challenges for both the implementer of a mixed-signal simulator and the model writer. We have discussed some of these challenges and their theoretical foundation, and we have provided suggestions how the challenges can be overcome. We hope that the information is also useful for designers, as it may help to understand some simulation failures.

REFERENCES

[1] IEEE Standard VHDL Analog and Mixed-Signal Extensions, IEEE Std 1076.1-1999
[2] IEEE Standard VHDL Language Reference Manual, ANSI/IEEE Std 1076-1993

[3] J. Vlach, K. Singhal: *Computer Methods for Circuit Analysis and Design*, Van Nostrand Reinhold, 1983

[4] K.E. Brenan, S.L. Campbell, L.R. Petzold: *Numerical Solution of Initial-Value Problems in Differential-Algebraic Equations*, SIAM Classics in Applied Mathematics, 1996

[5] L. Petzold: *Differential-Algebraic Equations are not ODE's*, SIAM J.Sci.Stat.Comput., vol. 3, no. 3, Sept. 1982, pp. 367-384

[6] M. Vlach: *Mixed-Mode-Simulator Interface*, U.S. Patent No. 4,985,860, Jan. 15, 1991

[7] S.E. Mattson, M. Andersson, K.J. Åström: *Object Oriented Modeling and Simulation*, in CAD for Control Systems, D.A.Linkens (Ed.), Marcel Dekker, Inc., 1993

[8] Kenneth Bakalar, private communication, 1996

[9] C.W. Gear, H.H. Hsu, L. Petzold: *Differential-Algebraic Equations Revisited*, Proc. ODE meeting, Oberwolfach, 1981

[10] C. Moler: *Are we there yet? Zero crossing and event handling for differential equations*, Matlab News & Notes, 1997 Special Edition

[11] R.F. Sincovec, A.M. Erismann, E.L. Yip, M.A. Epton: *Analysis of Descriptor Systems Using Numerical Algorithms*, IEEE Trans.Autom.Control, vol. AU-26, no. 1, Feb. 1981, pp. 139-147

[12] J. Vlach, J.M. Wojciechowski, A. Opal: *Analysis of Nonlinear Networks with Inconsistent Initial Conditions*, IEEE Trans. Circ.Syst.I, vol. 42, no. 4, Apr. 1995, pp. 195-200

Chapter 2

MIXED-SIGNAL EXTENSIONS FOR SYSTEMC

Karsten Einwich[1], Peter Schwarz[1], Christoph Grimm[2], Klaus Waldschmidt[2]
[1]Fraunhofer IIS/EAS Dresden, [2]University Frankfurt

Abstract: SystemC supports a wide range of Models of Computation (MoC) and is very
               well suited for the design and refinement of HW/SW-systems from functional
               downto register transfer level. However, for a broad range of applications the
               digital parts and algorithms interact with analog parts and the continuous-time
               environment. Due to the complexity of this interactions and the dominance of
               the analog parts in respect to the system behavior, is it essential to consider the
               analog parts within the design process of an Analog and Mixed Signal System.
               Therefore simulation performance is very crucial - especially for the analog
               parts. Thus different and specialized analog simulators must be introduced to
               permit the use of the most efficient solver for the considered application and
               level of abstraction. In this paper we describe possible areas of application and
               formulate requirements for analog and mixed-signal extensions for SystemC.

Key words: SystemC-AMS, Analog and Mixed Signal, System design

1.   INTRODUCTION AND MOTIVATION

SystemC 2.0 [1] provides a very flexible methodology for the design and refinement of complex digital HW/SW-systems. This methodology is strongly influenced by the communication model introduced by Gajski [10]. In this methodology, *modules* which consists of other modules or algorithms implemented in methods communicates via *channels*. A set of methods for communication is specified in an *interface*. These methods are realized in the channel. Modules can call methods of a channel, and events in a channel can activate methods in a module connected to the channel. This concept is generic enough to describe systems using various models of computation, including static and dynamic multirate dataflow, Kahn process networks,

E. Villar and J. Mermet (eds.), System Specification and Design Languages, 19–28.
© 2003 *Kluwer Academic Publishers.*

communicating sequential processes, and discrete events. We call such systems *discrete systems*.

However, the discrete approach to modeling and simulation is not applicable to systems, whose behavior is specified by differential and algebraic equations. In the following, we call such systems *analog systems*. Note, that the equations can also be given implicitly by a electrical netlist, by a mechanical model description or by transfer functions, for example. For many classes of systems, the analog parts will become more dominant in respect of performance parameters, power consumption, silicon cost and yield. Analog systems are simulated by mathematical methods that compute a solution for the underlying set of equations. For simulation of analog systems, a simulator *("solver")* computes a solution of the differential and algebraic equations, using linear and non-linear equation solving algorithms and numerical integration techniques to compute signal values for a time interval. For combined analog/discrete ("mixed-signal") simulation, discrete event simulators are coupled and synchronized with analog solvers.

In the following, we give an overview of the analog and mixed-signal extensions for SystemC discussed within the "SystemC-AMS working group" [5] (AMS=Analog and Mixed-Signal). First approaches for analog extensions are described in [2-4]. The synchronization mechanisms discussed in this paper are well known [6-9]. The main focus of this paper is to give an overview of the requirements for AMS extensions for SystemC and first approaches for their integration in SystemC:

– Which concept for AMS extensions corresponds well to the levels of abstraction, on which SystemC is used ?
– How can AMS extensions be integrated in SystemC ?

In section 20, areas of application and the level of abstraction, on which AMS-extensions should be applied is described. Section 23 gives an overview of the concept and introduces the layered approach for the integration of the AMS extensions into SystemC.

2.   SYSTEMC-AMS: AREAS OF APPLICATION AND REQUIREMENTS

Analog and mixed-signal systems can be found in many applications. The requirements for modeling and simulation depend both on the area of application and the level of abstraction of the model. SystemC is used for system-level design tasks, such as the modeling and refinement of hardware/software – systems. In such a context, analog models are used most notably for the following tasks:

Executable Specification: Analog models are often used as an executable specification of signal processing functions. Currently, interactive tools with a graphical or textual interface such as Matlab/Simulink are used for such tasks.

Behavioral Modeling: In the design of analog systems, there is always a large "bottom-up" fraction. Behavioral modeling of existing analog netlists allows the simulation of analog circuits in a reasonable time.

Co-Simulation with Environment: On system level, the analog (continuous-time) environment is co-simulated with the embedded system. This allows a rough validation of an executable specification. Furthermore, many designs can only validated in such a co-simulation.

In difference to digital systems, analog systems often combine different physical domains and are very application-specific. Therefore, on one hand, the use of analog and mixed-signal components in concrete applications must be considered. On the other hand, the approach must still be open for new applications and methodologies. For the discussion of application-specific requirements, three application fields are discussed: signal processing, RF/wireless and automotive applications.

In telecommunication and multimedia applications, the modeling of signal processing functions is dominant. These systems are mostly sampled or even over sampled using constant time steps. The system is modeled by a block diagram with directed signal flow. The blocks are described by linear transfer functions and weak or static non-linear functions. Often, such a system level description is combined with linear networks, which are used for macro modeling or for modeling the system environment.

In RF and wireless communication applications, systems are also specified by block diagrams with directed signal flow. A necessary feature for RF applications is the ability to simulate the baseband behaviour.

In the automotive domain, analog and mixed-signal systems are often non-linear systems, and usually embrace multiple physical domains (electrial, mechanical, fluidic, etc.). In difference to telecommunication and multimedia applications, control systems in the automotive domain are often systems with very different time constants ("stiff systems"). Nevertheless, executable specifications and executable prototypes are also modeled as block diagrams with directed signal flow.

The above mentioned requirements are partially contradictory: On the one hand, rather simple block diagrams connected with directed signal flow and some external components seem to be sufficient for many applications on a high level of abstraction. On the other hand, some requirements can only be fulfilled by solutions, which are specific for concrete applications and levels of abstractions. For example, for the simulation of electronic circuits a designer might want to use dedicated circuit simulators, such as

SABER or SPICE, and for the precise simulation of mechanical components a dedicated simulator for mechanical systems, and so on.

The situation discussed above leads us to an *open approach*. In such an approach, SystemC-AMS extensions are used for the simulation of executable specifications, behavioral models and the environment as far as possible. Furthermore, the AMS extensions provide a synchronization mechanism that permits the easy integration of additional solvers. These additional solvers can be specific for (maybe new) applications and levels of abstractions that are not covered by the "built-in" solvers. The resulting scenario is shown in Figure 1.

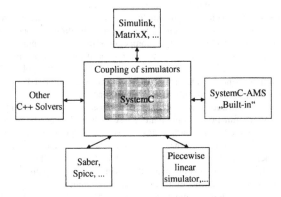

Figure 1: Use of SystemC-AMS in a heterogeneous tool-environment.

In addition to the requirements derived from the applications, AMS extensions for SystemC depend on the design and existing features of SystemC. SystemC already provides a generic and flexible concept for the modeling of computation and communication in discrete, heterogeneous systems. The existing SystemC is based on a layered structure of a C++ class library. Basically, SystemC consists of a generic discrete event simulator kernel, data types for modeling digital hardware, and templates for modeling communication and synchronization in different discrete models of communication (MoCs). In order to make communication and synchronization flexible, signals and ports use (communication) interfaces. For this reason, SystemC is more than just a discrete-event simulator. SystemC allows users to add new models of computation, as long as this is feasible with the discrete kernel. Such additional models of computation can be provided by libraries, for example, which define new classes of signals. The good object-oriented design of SystemC even invites users to make such extensions by simple inheritance. In this context, mixed-signal extensions for SystemC should provide a basic, object-oriented framework, where users can integrate appropriate and domain-specific "analog MoCs" – and not only an analog simulator which is coupled with the digital SystemC kernel. The mixed signal extensions shall allow users the integration of new solvers or alternative synchronization schemes in an as generic and flexible way as in the discrete domain. To allow the user this flexibility, a well-designed object oriented structure of the mixed-signal extension library will be required. In order to meet the different requirements, the AMS extension library is structured in four different layers. The structure of this "layered approach" is shown in Figure 2.

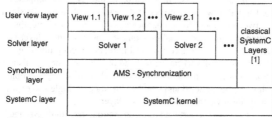

Figure: 2 Layered approach for SystemC-AMS extensions [5].

On top of the existing standard SystemC kernel, a *synchronization layer* provides methods to couple and synchronize different analog solvers and the discrete event kernel of SystemC. All communication between different MoCs takes place via this synchronization layer. On top of the synchronization layer, different analog MoCs can be realized by different analog solvers (*solver layer*) and the discrete event SystemC kernel. The *view layer* provides convenient user interfaces for the analog solvers. The view layer converts user views, for example netlists, to a set of differential or algebraic equations in the solver layer. The interaction of solver- and view layer can be compared to a model/view architecture, where one object provides an abstract data structure, another objects provide graphical views to the data structure.

3. THE LAYERED APPROACH IN DETAIL

The most critical part of the AMS extensions is the synchronization layer. The synchronization layer encapsulates the analog solvers as far as possible, and allows the connection of blocks simulated by different solvers or instances of the same solver for IP (intellectual property) encapsulation. In general, the synchronization layer handles the following tasks:

- The synchronization layer determines a step width for the simulated time, respecively the points in time, at which the solvers and the discrete event kernel are synchronized.
- The synchronization layer determines an order, in which the single solvers simulate.
- Cyclic dependencies can represent additional equations, which have to be solved. The synchronization layer handles this.

The concrete realization of the above functionality is rather application-specific. In order to limit the complexity, we make some restrictions:

1. For the communication via the synchronization layer, we suppose a directed signal flow. In the considered areas of application, the signal flow between system-level components is directed. Therefore, the realization of a directed signal flow on the synchronisation layer is sufficient. Note that - if necessary - a bidirectional coupling (e. g. for a coupling on netlist-level) can be realized by two separate signals in both directions. Furthermore, netlists or systems that are strongly coupled can (and should!) be modeled and simulated using one single solver without communication via the synchronization layer.

2. We do not plan to support backtracking (setting the simulated time back to the past). This would require the ability to save a state and to restore the saved state. We cannot expect solvers in general to support such methods. For this reason, the synchronization layer must only increase the simulated time.

In the telecommunications or RF domain, digital signal processing functions are usually oversampled with known and constant time steps. Therefore, a synchronization of the analog solvers and the digital kernel in constant time steps is sufficient. The order in which the solvers are computed is usually determined by the signal flow. This allows us to determine the order of execution of the induvidual simulators before the simulation. The underlying synchronization principle is the static dataflow model of computation.

In the automotive domain, systems are often nonlinear and stiff systems. The use of a constant step width could lead either to a large error, or to a very bad simulation performance. Furthermore, the coupling of analog and digital components is often very close. Therefore, the step width has to be determined during simulation by the synchronisation layer, controlled by parameters such as the local quantization error and/or synchronization requests and events of single simulators.

In order to provide a maximum of flexibility, a generalized, abstract "coordinator interface" provides access to the synchronization layer. This interface is by default realized by the rather simple, but stable static data-flow approach. In order to realize more complex synchronization schemes, the interface methods can be replaced by other, more sophisticated methods.

3.1   Simulation with the default method: Static dataflow

In order to determine an appropriate order of execution, the methods known from static dataflow MoC can be used. In the static dataflow MoC, cyclic dependencies are transformed into a pure linear dependency by assuming delays in each cycle as shown in Figure 3. The static dataflow MoC defines the order, in which the analog modules are simulated (1-4).

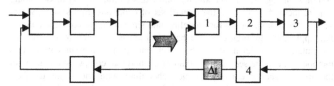

Figure 3: Synchronization of four analog modules with the static dataflow MoC.

However, the static dataflow MoC is an untimed model - the delay introduces to break the cyclic dependency has no numerical value. In order to simulate analog systems, time is mandantory. Therefore, we have to determine a concrete value for the introduced delay. A simple method is to ask each solver for the next point in the simulated time, when a synchronisation is required, and to select the minimum of all responses for all solvers. Then this value determines the value of the delay Δt. Furthermore, a user can specify a minimum and a maximum step width of simulators.

The above mentioned method for synchronisation looks very simple, but fulfills the requirements: At runtime, no scheduling effort has to be done, which makes computation of large structures with a large data volume possible, for example in telecommunication applications. All simulators can define synchronisation points, and no simulator is required to support backtracking. This allows the integration of very different solvers for the simulation of heterogeneous systems.

3.2   Possible improvements of the synchronisation methods

The default methods for synchronisation are rather general methods that permit the integration of nearly all solvers. Nevertheless, a user might need a faster method that reduces the number of points for synchronisation, maybe on the cost of simulation precision or with the risk of making errors – whatever is appropriate for his application. In order to reduce the number of synchronisations required, a single solver can be allowed to simulate for larger time steps than the step width chosen by the synchronisation layer. In this case, it might be necessary to *locally* do a backtracking within this solver. Note, that this backtracking is only local backtracing within one solver.

Other possible improvements are related to the way, in which the equations introduced by the structure of simulators are solved. When several analog models are coupled, the overall system can be seen as a more complex analog model. The simulation of this overall analog model would require methods that solve the overall set of equations. In a system, where very different approaches for modeling and simulation are combined, this is not applicable. In Figure 3, we have introduced a delay to explain the computational model used for the synchronization of the simulators. This method can also be seen as a relaxation method that computes a solution of the overall set of equations (The scheduling with the static dataflow approach can be seen as a sorting of the matrix according to the dataflow dependencies). Stability and convergence can then be improved by known methods that improve stability and convergence for relaxation, for example by a method that extrapolates a possible future waveform from the known waveform. In general, the synchronization layer provides an open platform for programmers and experienced designers to include alternative solvers and maybe alternative synchronization methods.

4.   SOLVER LAYER AND USER VIEW LAYER

A solver provides methods that simulate an analog system, e. g. a solver for a linear DAE-system, or another solver for a non-linear system, or a solver optimized for power electronics. A solver must be able to interact with the synchronization layer. For interaction with the synchronisation layer, he must provide at least methods that:
–   determine the next required synchronisation point, if a variable step width synchronization is required.
–   resume simulation with a given input waveform to a specified point in the simulated time.

The restrictions made in the synchronization layer lead to the following restrictions for the solvers:

- A solver may not set the simulated time of the overall simulation to a point in the past, because the synchronization layer does not support backtracking.
- A solver must accept synchronisation points from other simulators the synchronisation layer resp., because the synchronization layer determines the synchronization points.

Solvers simulate models that are specified by differential and algebraic equation. Those systems of equation are usually represented in a solver specific form, e.g. as constant or time dependent matrices. Examples for solvers that fulfill the above requirements are presented in [2, 3]. However, usually designers do not specify systems in such rather solver-specific data structures. A designer should describe systems using the *(user) view layer*. The view layer provides the solvers with the above-mentioned matrices. This representation of a system of equation can be generated from a network using the modified nodal analysis, or from a behavioral representation like transfer function or state space equations, for example. The same view can be useful for different solvers (e.g. linear/nonlinear DAE's); nevertheless, the realization must take into account that the mapping to the solver layer is different. At least the following views should be part of the SystemC-AMS extension library:

- *Netlist view:* This view should be common to all underlying solvers.
- *Equation view:* This view should allow a user to formulate behavioral models or functional specifications in a direct way as a differential algebraic equation.

Further views could be a possibility to describe finite element systems, for example – in general, the library should allow the user to add new and application specific views to a solver.

5. CONCLUSION

We have discussed requirements for analog and mixed-signal extensions for SystemC in a broad range of applications, for example in telecommunication, multimedia and automotive applications. The layered approach covers nearly all requirements for system-level design in these applications and can be extended in an easy way. The layered approach structures the extensions into a synchronization layer, a solver layer and a view layer.

Considering the requirements on system level, an open, general, but maybe less precise synchronization concept is more appropriate than a complicated precise, but closed concept. With the possibility to overload single methods of the synchronisation layer, the synchronization layer can be changed to more efficient synchronization methods – if required. The possibility to add solvers or external simulators in the solver layer allows us to include application-specific simulators in an easy and convenient way.

The proposed structure and concept of an analog mixed-signal extension library can significantly improve the generality of the SystemC approach: SystemC is currently restricted to the design of pure discrete systems. The proposed AMS extensions can make it "open" for the design of heterogeneous systems.

6.   REFERENCES

[1] An Introduction to System-Level Modeling in SystemC 2.0. Technical report of the Open SystemC Initiative, 2001. http://www.systemc.org/technical_papers.html.

[2] Karsten Einwich, Christoph Clauss, Gerhard Noessing, Peter Schwarz, and Herbert Zojer: "SystemC Extensions for Mixed-Signal System Design". Proceedings of the Forum on Design Languages (FDL'01), Lyon, France, September 2001.

[3] Christoph Grimm, Peter Oehler, Christian Meise, Klaus Waldschmidt, and Wolfgang Fey. "AnalogSL: A Library for Modeling Analog Power Drivers with C++. In Proceedings of the Forum on Design Languages", Lyon, France, September 2001.

[4] Thomas E. Bonnerud, Bjornar Hernes, and Trond Ytterdal: "A Mixed-Signal, Functional Level Simulation Framework Based on SystemC System-on-a Chip Applications". Proceedings of the 2001 Custom Integrated Circuts Conference, San Diego, May 2001. IEEE Computer Society Press.

[5] K. Einwich, Ch. Grimm, A. Vachoux, N. Martinez-Madrid, F. R. Moreno, Ch. Meise: "Analog Mixed Signal Extensions for SystemC". White paper of the OSCI SystemC-AMS Working Group.

[6] L. Schwoerer, M. Lück, H. Schröder: "Integration of VHDL into a Design Environment", in Proc. Euro-DAC 1995, Brighton, England, September 1995.

[7] M. Bechtold, T. Leyendecker, I. Wich: "A Dynamic Framework for Simulator Tool Integration", Proceedings of the 2nd International Workshop on Electronic Design Automation Frameworks, Charlottesville, 1990.

[8] M. Bechtold, T. Leyendecker, M. Niemeyer, A. Ocko, C. Ocko: "Das Simulatorkopplungsprojekt", Informatik-Fachbericht 255, Springer Verlag, Berlin.

[9] G. Nössing, K. Einwich, C. Clauss, P. Schwarz: "SystemC and Mixed-Signal – Simulation Concepts", in Proc. 4th European SystemC Users Group Meeting, Copenhagen, Denmark, October 2001.

[10] D. D. Gajski, J. Zhu, R. Dömer, A. Gerstlauer, S. Zhao: "SpecC Specification Language and Methodology", Kluwer Academic Publisher 2000.

Chapter 3

MIXED-DOMAIN MODELING IN MODELICA

Christoph Clauss[1], Hilding Elmqvist[3], Sven Erik Mattsson[3], Martin Otter[2], Peter Schwarz[1]

[1] *Fraunhofer Institute for Integrated Circuits, Branch Lab Design Automation, Dresden, Germany,* [2]*German Aerospace Center (DLR), Oberpfaffenhofen, Germany,* [3]*Dynasim AB, Lund, Sweden*

ABSTRACT   Modelica is a language for convenient, component oriented modeling of physical systems. In this article, an overview of the language, available libraries, the Dymola simulator and some industrial applications is given. Additionally, a comparison of Modelica with VHDL-AMS is presented.

1. INTRODUCTION

Modelica® is a non-proprietary specification of an object-oriented language for modeling of large, complex, and heterogeneous physical systems. It is suited for multi-domain modeling, for example, mechatronic models in robotics, automotive and aerospace applications involving mechanical, electrical, hydraulic and control subsystems, process oriented applications, and generation and distribution of electric power.

Models in Modelica are mathematically described by *differential, algebraic* and *discrete equations*. The set of equations does not need to be manually solved for a particular variable. A Modelica tool will have enough information to decide that automatically. Modelica is designed such that available, structural and symbolic algorithms can be utilized to enable efficient handling of large models having more than hundred thousand equations. Modelica is suited and used for hardware-in-the-loop simulations.

The Modelica design effort was started in September 1996. In January 2002, version 2.0 was released. More information on Modelica is available from http://www.Modelica.org.

E. Villar and J. Mermet (eds.), System Specification and Design Languages, 29–40.
© 2003 *Kluwer Academic Publishers.*

This paper is a very shortened version of the full paper presented at FDL'02. The full version is also available from http://www.Modelica.org.

2.   MODELICA FUNDAMENTALS

Modelica supports both high level modeling by composition and detailed library component modeling by equations. Models of standard components are typically available in model libraries. Using a graphical model editor, a model can be defined by drawing a *composition diagram* (also called schematics) by positioning icons that represent the models of the components, drawing connections and giving parameter values in dialogue boxes. Constructs for including graphical annotations in Modelica make icons and composition diagrams portable between different tools.

An example of a composition diagram of a simple motor drive system is shown in Figure 1. The system can be broken up into a set of connected components: an electrical motor, a gearbox, a load and a control system. The textual representation of this Modelica model is (annotations describing the graphical placement of the components and connection lines are not shown):

```
model MotorDrive
   PID       controller;
   Motor     motor;
   Gearbox   gear    (n=100);
   Inertia   inertia(J=10);
equation
   connect(controller.outPort, motor.inPort);
   connect(controller.inPort2, motor.outPort);
   connect(gear.flange_a    , motor.flange_b);
   connect(gear.flange_b     , inertia.flange_a);
end MotorDrive;
```

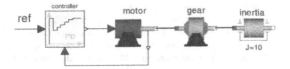

Figure 1: A model of a simple motor drive system.

It is a composite model, which specifies the topology of the system to be modeled in terms of components and connections between the components. The statement "Gearbox gear(n=100);" declares a component gear of model class Gearbox and sets the value of the gear ratio, n, to 100. A component model may be a composite model to support hierarchical modeling. The composition diagram of the model class Motor is shown in

Figure 2. The meaning of connections will be discussed below as well as the description of behavior on the lowest level using mathematical equations.

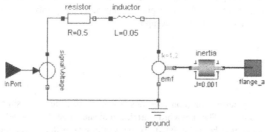

Figure 2: The model Motor.

Physical modeling deals with the specification of relations between physical quantities. For the drive system, quantities such as angle and torque are of interest. Their types are declared in Modelica as

```
type Angle  = Real(quantity="Angle",  unit="rad",displayUnit="deg");
type Torque = Real(quantity="Torque", unit="N.m");
```

where `Real` is a predefined type, which has a set of attributes such as name of quantity, unit of measure, default display unit for input and output, minimum, maximum, nominal and initial value.

Connections specify interactions between components and are represented graphically as lines between connectors. A connector should contain all quantities needed to describe the interaction. Voltage and current are needed for electrical components. Angle and torque are needed for drive train elements:

```
connector Pin            connector Flange
  Voltage      v;          Angle       phi;
  flow Current i;          flow Torque tau;
end Pin;                 end Flange;
```

A connection, `connect(Pin1, Pin2)`, with `Pin1` and `Pin2` of connector class `Pin`, connects the two pins such that they form one node. This implies two equations, `Pin1.v = Pin2.v` and `Pin1.i + Pin2.i = 0`. The first equation indicates that the voltages on both branches connected together are the same, and the second corresponds to Kirchhoff's current law saying that the current sums to zero at a node. Similar laws apply to mass flow rates in piping networks and to forces and torques in mechanical systems. The sum-to-zero equations are generated when the prefix `flow` is used in the connector declarations. In order to promote compatibility between different libraries, the Modelica Standard Library includes also connector definitions in different domains.

An important feature for building reusable descriptions is to define and reuse **partial** models. A common property of many electrical components is that they have two pins. This means that it is useful to define an interface model class OnePort, that has two pins, p and n, and a quantity, v, that defines the voltage drop across the component.

```
partial model OnePort
  Pin       p, n;
  Voltage v;
equation
  v = p.v - n.v;
  0 = p.i + n.i;
end OnePort;
```

The equations define common relations between quantities of a simple electrical component. The keyword **partial** indicates that the model is incomplete and cannot be instantiated. To be useful, a constitutive equation must be added. A model for a resistor extends OnePort by adding a parameter for the resistance and Ohm's law to define the behavior.

```
model Resistor "Ideal resistor"
  extends OnePort;
  parameter Resistance R;
equation
  R*p.i = v;
end Resistor;
```

A string between the name of a class and its body is treated as a comment attribute. Tools may display this documentation in special ways. The keyword parameter specifies that the quantity is constant during a simulation experiment, but can change values between experiments.

The most basic Modelica language elements have been presented. Modelica additionally supports mathematical functions, calling of external C and FORTRAN functions, control constructs, hierarchical data structures (record), replaceable components, safe definition of global variables and arrays, utilizing a Matlab like syntax. The elements of arrays may be of the basic data types (Real, Integer, Boolean, String) or in general component models, such as relays, switches, bearing friction, clutches, brakes, impact, sampled data systems, automatic gearboxes. This allows convenient description of simple, discretized partial differential equations. A unique feature of Modelica is the handling of discontinuous and variable structure [5]. Modelica has special language constructs allowing a simulator to introduce efficient handling of events needed in such cases. Special design emphasis was given to synchronization and propagation of events and the possibility to find consistent restarting conditions. Finally, a powerful package concept, similar to the Java package concept, is available to structure large model libraries and to find a component in a file system giving its hierarchical Modelica class name.

3.   MODELICA AND VHDL-AMS

Another well established modeling language for large, complex, and heterogeneous systems is VHDL-AMS (Very high speed integrated circuit Hardware Description Language – Analog and Mixed-Signal extensions) [1,11]. For categorizing both languages, some similarities and differences between VHDL-AMS and Modelica are shown next.

Using a spring model (Figure 3) with a linear characteristic a first impression of both languages is given. Both languages clearly distinguish between *interface* description and *internal* structural or behavioral description. This allows the description of complex models by the composition of simpler partial models.

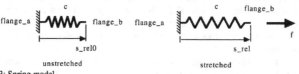

Figure 3: Spring model.

```
-- VHDL-AMS
ENTITY spring IS
  GENERIC
  (s_rel0: real := 0.0,
   c: real := 1.0);
  PORT
  (TERMINAL flange_a,
   flange_b: kinematic);
END ENTITY spring;

ARCHITECTURE
  simple OF spring IS
  QUANTITY s_rel ACROSS f
  THROUGH
    flange_b TO flange_a;
BEGIN
  f==c*(s_rel - s_rel0);
END ARCHTITECTURE simple;
```

```
// Modelica
model Spring
  import
    SI=Modelica.SIunits;
  Flange flange_a, flange_b;
  parameter
    Real c(min=0)= 1;
  parameter
    SI.Distance s_rel0 = 0;
    SI.Distance s_rel;
equation
  0=flange_a.f+flange_b.f;
  s_rel=
    flange_b.s-flange_a.s;
  flange_b.f =
    c*(s_rel-s_rel0);
end Spring;
```

The *structural* description with VHDL-AMS is a device oriented (branch oriented) netlist: it is noted which *nodes* and which *component ports* are connected together. In Modelica the connection of pins is described and nodes are not used explicitly.

Both languages support multi-domain and mixed-mode modeling. Designed for use in electronics, VHDL-AMS supports all necessary simulation modes used in electronic applications. Modelica has a slightly different philosophy: Before any operation is carried out, initialization takes place to compute a consistent initial point. Based on this initial point *transient analysis* or *linearization* can be carried out. The linearized model may be used to perform linear analysis calculations, like eigenvalue, or frequency response.

Modelica extensively allows object oriented features like the definition of classes, inheritance and polymorphism, whereas VHDL-AMS uses overloading of operators. A key feature of Modelica is its library concept (package concept), which uses ideas from the Java programming language whereas VHDL-AMS uses the package concept of the programming language Ada.

A VHDL-AMS model description is purely textual and does not include graphical information. The model text may be generated by powerful schematic entry tools in design frameworks. As a consequence, an exchange of VHDL-AMS models between different simulators is portable only on the textual level. In Modelica all details of the graphical layout of a composition diagram on screen and on printer is standardized

Further differences exist with regard to the analog-digital simulation algorithm:

VHDL-AMS presupposes the coupling of a discrete event-driven algorithm with a continuous DAE solver, the coupling algorithm is part of the language standard. The analog solver has to determine the actual values of all analog quantities at each event time. The digital signals are calculated by a discrete event-driven algorithm that is part of the VHDL standard. Digital events occur only at time points of a discrete time axis.

In Modelica at all time instants, including events, the whole set of active differential, algebraic and discrete equations is solved as one (algebraic) system of equations, at least conceptually [5]. The advantage is that the synchronization between the continuous and the discrete part is automatically determined at translation time by data flow analysis (= sorting of the equations). This allows, e.g., to figure out non-deterministic behavior before simulation starts. For the formulation of analog-digital conversion both languages have special constructs (VHDL-AMS: 'ramp, 'above, 'slew, break; Modelica: **pre**(..), **edge**(..), **change**(..), **noEvent**(..), if-then-else conditions).

In the design of electronic systems and in the simulation of their environments, VHDL-AMS offers many advantages, e.g., powerful digital and mixed-signal simulators and large model libraries. Many IC companies require checked VHDL (or Verilog) simulation protocols as part of their quality assurance system. In non-electrical domains, especially with a large part of continuously working subsystems, Modelica offers advantages by its object orientation, handling of complicated modeling situations, the graphic-oriented input language, and many multi-domain libraries.

4.   MODELICA LIBRARIES

In order that Modelica is useful for *model exchange*, it is important that libraries of the most commonly used components are available, ready to use, and sharable between applications. For this reason, the Modelica Association develops and maintains a growing *Modelica Standard Library*. Furthermore, other people and organizations are developing free and commercial Modelica libraries. Currently, component libraries are available in the following domains:

- About 450 type definitions, such as Angle, Voltage, Inertia
- Mathematical functions such as sin, cos, ln
- Continuous and discrete input/output blocks (transfer functions, filters...)
- Electric and electronic components
- 1-dim. translational components such as mass, spring, stop
- 1-dim. rotational components (inertia, gearbox, planetary gear, clutch...)
- 3-dim. mechanical components such as joints, bodies and 3-dim. springs.
- Hydraulic components, such as pumps, cylinders, valves
- 1-dim. thermal components (heat capacitance and thermal conductor...)
- Thermo-fluid flow components (pipes, heat exchangers ...)
- Building components (boiler, heating pipes, valves, weather conditions..)
- Power system components such as generators and lines
- Vehicle power train components such as driver, engine, gearboxes
- Flight dynamics components (bodies, aerodynamics, engines, wind ...)
- Petri nets, such as place, simple transition, parallel transition

ELECTRICAL SUBLIBRARY

In electronics and especially in micro-electronics, the simulator SPICE [4] with its input language is a quasi-standard. SPICE has only a fixed set of device models ("primitives"). These models can be modified by changing their parameters. The transistor models are very sophisticated and are closely integrated into the SPICE simulator. Behavioral modeling by the user is not supported. To model complicated functions, "macromodels" in the form of circuits have to be constructed which use the SPICE primitives as building blocks in a netlist.

The Modelica.Electrical.Analog library has been developed to support modeling of electrical subsystems. It is a collection of simple electronic device models, independent of SPICE. They are easy to understand, use the object-oriented features of the Modelica language, and can be used as a base of inheritance and as examples for new models. The functionality is chosen in a way that a wide range of simple electric and electronic devices can be modeled. The Modelica.Electrical.Analog library contains at the moment:

- Basic devices like resistor, capacitor, inductor, transformer, gyrator, linear controlled sources, operational amplifiers (opamps).
- Ideal devices like switches (simple, intermediate, commuting), diode, opamp, transformer.
- Transmission lines: distributed RC and RLGC lines, loss-less transmission lines.
- Semiconductors: diodes, MOS and bipolar transistors (Ebers-Moll model).
- Various voltage and current sources.

The numerically critical devices of the library are modeled in a "simulator-friendly" way, e.g. exponential functions are linearized for large argument values to avoid data overflow.

In future, the SPICE models have still to be provided. An interesting approach is presented in [10], where the SPICE models are extracted and translated into Modelica. Another possibility is the call of SPICE models via the foreign function interface. Furthermore, a library of *digital* electronic devices and of composed components, e.g., CMOS building blocks, is under development.

5.   THE SIMULATION PROBLEM

In order that the Modelica language and the Modelica libraries can be utilized, a Modelica translator is needed to transform a Modelica model into a form, which can be efficiently simulated in an appropriate simulation environment. The mathematical simulation problem consists of the equations of all components and the equations due to connections. It is a hybrid system of differential-algebraic equations (DAE) and discrete equations. Moreover, idealized elements, such as ideal diodes or Coulomb friction introduce *Boolean unknowns* (e.g., for friction models the Boolean variables describe whether the element is in stuck mode or is sliding). Thus the simulation problem may be a *mixed* system of equations having both *Real* and *Boolean unknowns*. There are no general-purpose solvers for such mixed problems.

Modelica was designed such that symbolic transformation algorithms can (and have to) be applied to transform the original set of equations into a form which can be integrated with standard methods. With appropriate symbolic transformation algorithms such as BLT-partitioning (= sorting of equations and variables) and tearing (= "intelligent" variable substitution), it is possible to reduce the number of unknowns visible from the integrators during translation.

When solving an ordinary differential equation (ODE) the problem is to integrate, i.e., to calculate the states when the derivatives are given. Solving a DAE may also include differentiation, i.e., calculate the derivatives of given variables. Such a DAE is said to have high index. It means that the overall number of states of the model is less than the sum of the states of the sub-components. Higher index DAEs are typically obtained because of constraints between models.

Therefore, any reliable Modelica translator must transform the problem before a numerical solution can take place. The standard technique is to (1) use the algorithm of Pantelides [9] to determine how many times each equation has to be differentiated, (2) differentiate the equations analytically, and (3) select which variables to use as state variables (either statically during translation or in more complicated cases dynamically during simulation) with the dummy derivative method of Mattsson and Söderlind [7].

6. THE DYMOLA SIMULATION ENVIRONMENT

Dymola [2] from Dynasim (http://www.Dynasim.se) has a Modelica translator, which is able to perform all necessary symbolic transformations for large systems (> 100 000 equations) as well as for real time applications. A graphical editor for model editing and browsing, as well as a simulation environment are included.

The traditional approach has been to manually manipulate the model equations to ODE form. Dymola automates all this time-consuming and error-prone work and generates efficient code also for large and complex models. Symbolic processing is a unique feature of Dymola to make simulations efficient. Dymola converts the differential-algebraic system of equations symbolically to state-space forms, i.e. solves for the derivatives. Efficient graph-theoretical algorithms are used to determine which variables to solve for in each equation and to find minimal systems of equations using tearing to be solved simultaneously (algebraic loops). The equations are then, if possible, solved symbolically or code for efficient numeric solution is generated. Discontinuous equations are properly handled by translation to discrete events as required by numerical integration routines.

For high index problems, Dymola deduces automatically which equations to differentiate and differentiates them symbolically. It also selects which variables to use as states.

Often one challenge is real-time simulation for hardware-in-the-loop testing. For off-line simulations it is interesting to reach the stop time as fast as possible. Stiff solvers may spend a lot of effort to calculate transients but have the ability to speed up once the fast transients have died out. In real-time simulations it is necessary to produce values at a certain sampling rate and it is the worst case that sets the limit for speed.

When using the explicit Euler method for solving the ODE the stepsize is typically limited due to stability. Often the problems are stiff which causes such a small stepsize that real-time simulation is not possible. Using the implicit Euler method implies that a nonlinear system of equations needs to be solved at every step. The size of this system is at least as large as the size of the state vector, n. Solving large nonlinear systems of equations in real-time is somewhat problematic because the number of operations is $O(n^3)$ and the number of iterations might vary for different steps. Reducing the size of the nonlinear problem is advantageous. Due to the hybrid nature of the system, the Jacobian of the nonlinear system can change drastically between steps. This makes it difficult to apply methods relying on Jacobian updating.

In order to obtain real-time capability special symbolic and numerical model processing and simulation code generation methods have been developed for Dymola. The method of inline integration [3] was introduced to handle such cases. The discretisation formulas of the integration method are combined with the model equations and structural analysis and computer algebra methods are applied on the augmented system of equations. E.g., for a robotic model with 66 states, Dymola reduces the size of the nonlinear system to only 6 with no additional local equation systems. The equation systems from discretizing the drive trains and their controllers are linear and Dymola is able to solve them symbolically.

Convenient interfaces to Matlab and the popular block diagram simulator SIMULINK exist. For example, a Modelica model can be transformed into a SIMULINK S-function C mex-file, which can be simulated in SIMULINK as an input/output block.

7.   APPLICATION: HARDWARE-IN-THE-LOOP SIMULATION OF AUTOMATIC GEARBOXES

Hardware-in-the-loop simulation of automatic gearboxes to tune gearshift quality is a major application of Modelica and Dymola for automotive companies. A gearbox is modeled by using predefined components see Figure 4 for a Modelica composition.

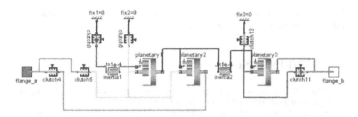

Figure 4: Modelica composition diagram of automatic gearbox model.

A typical electronic control unit, ECU, which generates the gearshift signals has a sampling time of 5 - 10 ms implying that the simulation needs to produce values in this time frame or faster. A model is built by composing planetary wheel-sets, shafts, clutches, and free-wheels from the Modelica libraries. The resulting mathematical model is a mixed system of Boolean equations and differential-algebraic equations.

The gearbox model has varying structure depending on whether a clutch is sliding or not. There are different combinations of bodies that are rotating together. For a given configuration the simulation problem is much simpler. The traditional approach has been to manually manipulate the equations for each mode of operation and exploit special features. Dymola automates all this work. It takes Dymola less than a minute to translate the Modelica model of the gearbox into efficient simulation code. Already in 1999, a 500 MHz DEC Alpha processor from dSPACE evaluated one Euler step including a possible mode switch in less than 0.18 ms. Today a modern PC processor is 3-4 times faster.

8.   CONCLUSION

The first useful Modelica environment, Dymola, has been available since beginning of year 2000. Since then an increasing number of modelers is using Modelica both in research and in industrial applications. The development of the Modelica language and of Modelica libraries is still an ongoing effort in order to improve the language and the libraries based on the gained experience. In future Modelica will probably also be more and more used not only for modeling but also for programming of algorithms, because it turned out that the combination of the Matlab-like array syntax in Modelica functions and the strong typing semantic seems to allow both

Matlab-type convenience of algorithm definition and C-type run-time efficiency of the translated algorithm.

9.   REFERENCES

[1] Christen, E.; Bakalar, K.: A hardware description language for analog and mixed signal applications. IEEE Trans. CAS-II 46 (1999) 10, 1263-1272.

[2] Dymola: Dynamic   Modeling   Laboratory.   Dynasim   AB,   Lund,   Sweden, http://www.Dynasim.se

[3] Elmqvist, H.; Otter, M.; Cellier, F.E.: Inline integration: A new mixed symbolic/numeric approach for solving differential-algebraic equation systems. European Simulation Multiconference June 1995, Prague, 23-34.

[4] Johnson, B.; Quarles, T.; Newton, A.R.; Pederson, D.O.; Sangiovanni-Vincentelli, A.: SPICE3 Version 3e, User's Manual. Univ. of California, Berkeley, Ca., 94720, 1991

[5] Mattsson, S.E.; Otter, M. Hilding, E.: Modelica Hybrid Modeling and Efficient Simulation. 38th IEEE Conference on Decision and Control, CDC'99, Phoenix, Arizona, USA, pp. 3502-3507, Dec 7-10, 1999.

[6] Mattsson, S.E.; Elmqvist, H.; Otter, M.; Olsson, H.: Initialization of hybrid differential-algebraic equations in Modelica 2. Modelica'2002, Oberpfaffenhofen, March 18-19, pp. 9-15, 2002.
   (download from http://www.Modelica.org/Conference2002/papers.shtml)

[7] Mattsson S.E.; Söderlind G.: Index reduction in differential-algebraic equations using dummy derivatives. SIAM Journal of Scientific and Statistical Computing, Vol. 14, pp. 677-692, 1993.

[8] Modelica Association: Modelica[®] - A Unified Object-Oriented Language for Physical Systems Modeling, Language Specification, Version 2.0.
   (download from http://www.Modelica.org/Documents/ModelicaSpec20.pdf)

[9] Pantelides C.: The Consistent Initialization of Differential-Algebraic Systems. SIAM Journal of Scientific and Statistical Computing, pp. 213-231, 1988.

[10] Urquia, A., Dormido, S.: DC, AC Small-Signal and Transient Analysis of Level 1 N-Channel MOSFET with Modelica. Modelica'2002, Oberpfaffenhofen, March 18-19, 2002.
   (download from http://www.Modelica.org/Conference2002/papers.shtml)

[11] VHDL-AMS: http://www.vhdl.analog.com (information of the IEEE 1076.1 working group)

Chapter 4

VHDL-AMS AND VERILOG-AMS AS COMPETITIVE SOLUTIONS
for the High Level Description of Thermoelectrical Interactions in Opto-Electronic Interconnection Schemes

François Pêcheux, Christophe Lallement
ERM-PHASE, ENSPS
Bld. S. Brant, Pôle API, 67400 Illkirch, France
{francois.pecheux,christophe.lallement}@ensps.u-strasbg.fr

Abstract: The paper details and compares two implementations of the same opto-electronic interconnection scheme, coded with Mixed Hardware Description Languages VHDL-AMS and Verilog-AMS. The scheme encompasses several interacting domains (thermal, optical and electrical). It includes a temperature sensitive transmission emitter designed with deep sub-micron technology transistors, a laser diode for light emission, a transmission medium (free space or waveguide) and a transmission receiver, composed of a single photodiode. First, we describe the hierarchical model itself. Second, we present the features common to VHDL-AMS [1,2] and Verilog-AMS [3,4,5]. Third, we describe the main parts of the model with these two HDLs, with a special emphasis on conservative/signal-flow semantics and model interconnections.

Key words: EKV 2.6 MOST Model, Thermo-electronic interaction, Opto-electronic interconnection, Verilog-AMS, VHDL-AMS

1. INTRODUCTION

The use of optics for signal transmission offers many attractive advantages over pure electronic approaches, and optical interconnects can now be found in small computer boards or single chips. A typical opto-electronic interconnection scheme includes a transmission emitter with its associated laser diode for light emission, a transmission medium and a transmission receiver, usually composed of a single phototransistor.

E. Villar and J. Mermet (eds.), System Specification and Design Languages, 41–50.
© 2003 *Kluwer Academic Publishers.*

Accurate modeling and simulation of such systems play a key role in the telecommunication industry. CAD tools exist today that enable the design and simulation of digital Systems-On-Chip (SOC), as well as optical simulators, but the two worlds have long been clearly separated. Until 1999, no consideration has been given to the optical properties of the laser diode when designing the transmitter electronics.

Moreover, thermal aspects have been neglected most of the time. With sub-micron or deep sub-micron transistor dimensions and the need for speed in today's microprocessor core, a key issue is the modeling of thermoelectrical interaction between devices in such systems. These thermal effects are constantly amplified by the growing power density, and a failure in their estimation at an early development stage of the design often means extra costs and delays.

Since 1999, new "System-oriented" CAD tools have emerged, and two unified answers to this key mixed domains modeling and simulation issue have been proposed. They both are based on innovative Mixed Hardware Description Languages: VHDL-AMS and Verilog-AMS. In this paper, we propose to discuss the possibilities and the major syntactic constructs of the two languages, with a special emphasis on signal connections semantics and thermodynamical interactions between inner devices, and the limitations of the currently available tools.

2.   A TYPICAL OPTO-ELECTRONIC INTER-CONNECTION SCHEME

Figure 1 acts as a reference throughout the whole paper. It presents a complete interconnection system, composed of a transmission emitter, a transmission medium (free space or waveguide), and a receiver (photodiode). Parts in black of Figure 1 represent electrical interconnections, while light grey parts are related to thermal interactions.

The transmission emitter contains instances of an EKV N-MOST sub-model and a P-MOST sub-model connected as a CMOS inverter. The EPFL EKV version 2.6 MOST sub-model [6] is a charge-based compact model that consistently describes effects on charges, transcapacitances, drain current and tranconductances in all regions of operation of the MOSFET transistor (weak, moderate, strong inversion) as well as conduction to saturation. The effects modeled in this sub-model include all the essential effects present in sub-micron technologies. In the EKV MOST electrical description, several electrical parameters are highly dependent on temperature. Their respective temperature variations are taken into account by appropriate coefficients in the sub-model equations [6].

In Figure 1, the N-MOST and P-MOST sub-model instances are surrounded by 5 filled black circles that represent terminals. A terminal, according to the VHDL-AMS and Verilog-AMS terminologies, is a connection port with well-defined conservative semantics. Four of them (Drain, Gate, Source, Bulk) represent purely electrical connexions. The fifth MOST connector, TJ, is of type thermal and is used to model thermal exchanges. The laser diode is modeled as a four pins device [7]. Two of them are electrical terminals (Anode and Cathode), TJ is a thermal terminal. The fourth pin is a signal-flow connector, L$_{out}$, as shown by the white circle, and manages power light propagated by the transmission medium. The laser diode device is also very sensitive to temperature [7].

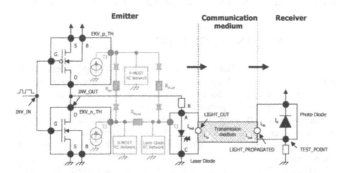

Figure 1. The studied thermo-opto-electronic transmission scheme.

The structure composed of the N-MOST, P-MOST and Laser Diode is temperature dependent. Heat diffusion through solid materials can be modeled by sourcing dissipated power of each of these elements into a thermal RC network (thermal resistances and capacitances) [8], which represents the material properties of the different layers from chip to ambient. Resistors R$_{NLed}$, R$_{PLed}$, and R$_{NP}$ of the transmission emitter model the relative coupling of each element with its other two neighbors.

The temperature profile [9,10] of the global model is the result of heat flow in the thermal network. In such networks, energy conservation states that the sum of all power contributions at a thermal node equals zero, and that the temperature at all terminals connected to a thermal node be identical. Thermal evolution of a system is thus ruled by the very same conservative Kirchhoff laws dictating the behavior of electrical systems: voltage is replaced by the across quantity temperature, and current becomes the through quantity heat flow.

Modeling the transmission medium without geometrical considerations and no dispersion means pure delay with a slight reducing factor. Light source is connected to the input, and is propagated to the output after a certain delay, function of distance and light phase speed. The connection ports of the transmission medium, L_{In} and L_{Out}, follow signal-flow semantics, i.e. the transmission medium has infinite input impedance and null output impedance. This explains why transmission line characteristics are not dependent on the loads applied to it, and hence why fiber optics are so widely used. The photodiode is a device that generates current when it receives light L_{In}. Its interface contains an input connector for light L_{In}, two electrical terminals (anode and cathode) and a thermal terminal dedicated to thermodynamical exchanges with other devices in the transmission receiver.

3.　　TWO LANGUAGES, ONE GOAL

The intent of both VHDL-AMS and Verilog-AMS is to let designers of analog and mixed signal systems and integrated circuits create and use module that encapsulate high-level behavioral descriptions as well as structural descriptions of systems and components.

The behavior of each module can be depicted in terms of its mathematical equations binding terminals and parameters, and the structure of each component can be described by a set of interconnected sub-components.

VHDL-AMS has been defined as IEEE standard IEEE 1076-1999. Unlike Verilog, Verilog-AMS is not a IEEE standard yet. VHDL-AMS and Verilog-AMS are directly competitive, but nethertheless share some common features:
- Differential-algebraic equations,
- Conservative or signal-flow systems,
- Mixed disciplines (mechanical, electrical, rotational ...) capability,
- Parameterization and hierarchy,
- Programming structures (loops, if-then-else, assignments),
- Analog operators (delay, transition, slew, noise),
- Time Derivative and Integral Operators,
- Linear continuous time filters with Laplace Transforms.

For the system designer, one of the major interest of these two languages is the simplicity with which models involving distinct physical domains (electronical, optical, mechanical, etc) can be interconnected, through conservative and signal flow assignments. Along with the ability to describe hierarchical/multi-abstraction models, the power of VHDL-AMS and Verilog-AMS relies on the fact that it allows the model designer to only detail the constitutive equations of the inner sub-models. These HDLs take advantage of a graph-based conceptual description to compute the equations describing the conservative Kirchhoff laws of the global model.

4. MODELING THERMAL-ELECTRICAL DEVICE: THE EKV MOST MODEL

The purpose of this part is to present simultaneously the VHDL-AMS and Verilog-AMS codes for a complex simulation model and to detail the mechanisms by which thermoelectrical interactions can take place with these two HDLs. Numbers like ❷ in the text reference bookmarks in the following code, and help understanding the code flow. Numbers do not necessarily appear in the same order in both codes, because of differences in the HDL structures. Listing 1 presents both codes for the EKV nMOST model.

4.1 Libraries (bookmark ❶)

Both VHDL-AMS and Verilog-AMS codes first contain references to libraries needed to parse the model. In VHDL-AMS, libraries defining disciplines like electrical and thermal are located in separate packages, while in Verilog-AMS everything is part of the disciplines.h file. Because the internal equations of EKV rely on smoothing functions, the math library is also included in VHDL-AMS. In Verilog-AMS, these transcendental functions can directly be accessed.

4.2 The EKV Interface (bookmark ❷)

For the sub-model end-user (circuit designer), the most important part is the interface, contained in what is called an entity in VHDL-AMS, and a module in the Verilog-AMS terminology. One can see that VHDL-AMS distinguishes between the model interface and the model body, containing the internal equations. The n-MOST EKV_n_TH entity contains the traditional 4 terminals, all of electrical type, plus a fifth thermal terminal, TJ, which allows heat exchanges between the transistor and its direct environment. All the terminals are part of a port statement.

In Verilog-AMS, the n-MOST EKV_n_TH module contains the same 5 terminals (d,g,s,b,tj), declared as inout. Verilog-AMS compliant tools do not currently support correct application of terminal directions.

4.3   Defining generic parameters (bookmark ❷)

In VHDL-AMS, the `generic` statement in the `entity` allows the designer to define parameters which values can be overridden during instanciation of the sub-model. Typically, geometrical `weff` and `leff` transistor parameter are defined as generic.

In Verilog-AMS, generic parameters are prefixed with the `parameter` token.

4.4   Accessing terminals (bookmark ❸)

Energy conservation states that the behavior of a sub-model depends on its direct neighbors connected to its terminals. Kirchhoff laws associate to each terminal of a given nature (electrical, thermal, etc) two important parameters, defined as intensive and extensive in general thermodynamics. The intensive parameter is an `across` parameter in VHDL-AMS and a `potential` parameter in Verilog-AMS. The extensive parameter is a `through` parameter in VHDL-AMS and a `flow` parameter in Verilog-AMS. To access one of these parameters from the terminals, VHDL-AMS and Verilog-AMS follow two distinct approaches.

In VHDL-AMS, access to terminals is done through the use of *bound* quantities. Quantities represent the static behavior of the continuous system, and correspond to the unknowns of the resulting set of ordinary differential algebraic equations that have to be solved simultaneously by the analog solver over time. Mathematically, the number of unkowns must match the number of equations, or else the algebraic system is over or underdetermined. Quantities that are not bound to any terminal are said to be *free*. Line ❸ states that bound quantity `vg` is defined as the potential between terminals `g` and `b`. Through this mechanism, any equation involving `vg` will contribute to compute the `g` and `b` terminal potential. Similarly, `gpower` is a through (flow) quantity, and `temp` is the corresponding potential.

In verilog-AMS, one can explicitly access the potential or flow parameter of a terminal through the use of *access functions*. When applied to a specific electrical terminal, `V()` or `I()` respectively give access to the voltage or current. This is a major difference with VHDL-AMS.

4.5   Analog modeling (bookmark ❹)

The first set of equations represents the thermal-dependent electrical parameters that are constantly reassigned according to temperature variation.

Once preprocessed, these parameters are applied in the analytical equations of the EKV electrical model.

In VHDL-AMS, the EKV model is implemented as a set of simultaneous statements between the `begin` and `end` keywords. At each Analog Simulation Point (ASP), all these equations are reevaluated.

VHDL-AMS Code	Verilog-AMS Code

⦿ `library disciplines;`
`use disciplines.electromagnetic_system.all;`
`use disciplines.thermal_system.all;`
`library ieee;`
`use ieee.math_real.all;`
 	
`entity EKV_n_TH is`
⦿ `generic`
`(`
`Weff : real := 1.5e-6; -- channel width`
`Leff : real := 0.15e-6; -- channel length`
`.../...`
`);`
⦿ `port (terminal d,g,s,b: electrical;`
`terminal tj:thermal);`
`end;`
`architecture equ of ekvn is`
 	
⦿ `quantity vg across g to b;`
`quantity vd across d to b;`
`quantity vs across s to b;`
`quantity id through d;`
`quantity isource through s;`
`quantity gpower through thermal_ground to tj;`
`quantity temp across tj to thermal_ground;`
 	
`constant boltz : real := 1.3806226e-23;`
`constant charge : real := 1.6021918e-19;`
`constant konq : real := boltz / charge;`
`.../...`
 	
⦿ `begin`
`vt == konq*temp+1.0e-20;`
`.../...`
`isource == -id;`
`gpower == abs(id * (vd-vs));`
`end;`
	⦿ `` `include "disciplines.h" ``
	
	⦿ `module EKV_n_TH(d,g,s,b,tj);`
	`// Node definitions (external nodes)`
	
	`inout d,g,s,b,j ;`
	`electrical d,g,s,b ;`
	`thermal tj ;`
	
	`//*** model parameter definitions`
	⦿ `parameter real boltz = 1.3806226E-23;`
	`parameter real charge = 1.6021918E-19;`
	`//** geometrical parameters`
	`parameter real weff = 100.0E-6;`
	`parameter real leff = 100.0E-6;`
	`.../...`
	
	`//*** Local variables`
	`real konq, vt;`
	`.../...`
	`real vg,vd,vs;`
	`real id,isource;`
	
	⦿ `analog begin // EKV v2.6 long-channel`
	`vg = V(g,b);`
	`vs = V(s,b);`
	`vd = V(d,b);`
	
	⦿ `konq =boltz/charge;`
	`vt = konq * Temp(j) + 1.0E-20;`
	`.../...`
	`isource = -id;`
	
	`I(d) <+ id;`
	`Pwr(j) <+ abs(id * (vd-vs));`
	`end // analog`
	`endmodule`

Listing 1. The Verilog-AMS and VHDL-AMS codes for a simplified version of the EKV N-MOST model, with thermal interactions.

In Verilog-AMS, the keyword `analog` initiates the behavioral modeling. Unknowns are declared as real floating-point types. Statements that appear into an `analog` block are in the continuous context but, unlike VHDL-AMS, are executed sequentially. Verilog-AMS distinguishes between constitutive relationships, that describe the analog behavior of the model, and interconnection relationships that describe the structure of the conservative graph. Former relationships are realized through simple assignments and the latter with the important source branch contribution operator `<+`. To make a sub-model a conservative system, the model designer has to use the access functions and the source branch contribution operator appropriately. Access functions to a potential or flow variable can be used in the right part of any ODAE of the sub-model but to assign a conservative variable (left part of an equation) like voltage or current of a node, the model designer has to use the <+ operator.

In the EKV MOST model, the drain current associated to terminal d receives a current contribution provided by the variable Id, assigned on the previous line. In the case of several analog blocks, all the contributions are summed up to establish I(d). The power associated to terminal j receives a power contribution through the access function Pwr() provided by the product of current Id by voltage bias (vd-vs). As an exotic add-on, one must known that thermodynamical voltage (vt in the equations) can directly be obtained in Verilog-AMS with the $vt function. However, for our purpose, we have not used this possibility.

Accordingly, once the analogy between electrical and thermal networks has been established, the modeling of thermal networks is straightforward.

Thus, the major difference between VHDL-AMS and Verilog-AMS relies on the way the conservative graph is built. On one hand, VHDL-AMS makes extensive use of a powerful simultaneous statement to model equations and the conservative graph is build implicitly during elaboration thanks to bound quantities. On the other hand, Verilog-AMS is based on an explicit description of the conservative graph through the use of the contribution operator. In a sense, VHDL-AMS is equation oriented, while Verilog-AMS is based on a nodal analysis scheme.

5.   MODELING OPTO-ELECTRICAL DEVICE: LED, TRANSMISSION MEDIUM, AND PHOTODIODE

A laser diode only emits light when it is applied a forward current greater than a threshold current [7]. From a syntactic point of view, it means that one equation out of two has to be chosen, given a certain condition on variables. In VHDL-AMS, this choice is managed by the if use construct. In Verilog-AMS, it is a simple if else statement, but a switch branch lies underneath ❶. Listing 2 shows both codes for the laser diode.

Moreover, the laser diode does not respond immediately when it is applied a forward current. Like any physical device, the laser physical effect takes some time to be established. The delay corresponds to a low pass-filter, with a time constant tau. Thanks to the time derivative and integral operator, the modeling of such a filter is easy, as shown in Listing 2 ❶. One could also have used the Laplace Transfer Function 'LTF with parameters.

Light emission by the laser diode is not subject to Generalized Kirchhoff laws. In VHDL-AMS, a signal-flow signal is nothing more than a quantity declared in the port section of the entity. Its associated direction defines whether it is communicated to the model (in) or propagated out of the model (out). In Verilog-AMS, quantity equivalents can't be declared in the interface. To have signal-flow capabilities associated to a terminal, one has to declare a new discipline light with no associated flow nature, or just use an existing discipline like Voltage and make the analogy. In this code, light is a new discipline that defines a new nature lightp considered as a potential parameter. The corresponding access function is LP(), standing for Light Power. For the transmission medium, in VDHL-AMS, delaying a quantity is done through the use of the 'delayed attribute. In Verilog-AMS, the delay operator has the same function. The photodiode is modeled by the very same means.

VHDL-AMS Code	Verilog-AMS Code
.../...	`include "disciplines.h"

ENTITY diode_laser IS
  GENERIC (Ir : real := 1.0e-12;
.../...
  );
  PORT (terminal inp,inm :electrical;
    terminal j:thermal ;
	module diode_laser(inp,
	inm,j,lightout) ;
	
	inout inp,inm,j,lightout ;
	electrical inp,inm ;
	thermal j ;

quantity lightOut : out real); END ENTITY diode_laser;	light lightout ;
ARCHITECTURE equ OF diode_laser IS .../...	parameter real ir = 1.0e-12; .../...
BEGIN	real id ; real vd ;

.../...
❶ If id>Ith_real use
  lightPower == eta_real*(id-Ith_real);
else
  lightPower == 0.0;
end use;
❷ Light == lightPower-tau*Light'dot;
lightOut == Light;
vt == k * temp / q;
LightOut == ...
.../..
power == id * vd;
 	
END ARCHITECTURE equ;
	real lightpower;
	real plight;
	
	analog
	begin
	.../...
	    vd = V(inp,inm) ;
	  ❶ if (I(inp,inm)>ith_real)
	      lightpower = eta_real *(id-ith_real);
	    else
	      lightpower = 0.0 ;
	  ❷ plight = lightpower-tau*ddt(plight) ;
	    LP(lightout) <+ plight ;
	.../...
	    I(inp,inm) <+ id;
	    Pwr(j) <+ id * vd ;
	  end
	endmodule

Listing 2. Modeling the laser diode, in VHDL-AMS and Verilog-AMS.

6. CONCLUSION

Simulation results extracted from the Anacad ADVanceMS VHDL-AMS simulator and complete code listings can be found in [11]. Both languages are good answers to the mixed modeling issue, and are natural extensions of the initial ideological rift between VHDL and Verilog adepts. They both provide the necessary mechanisms to handle interconnections of distinct domain models, and conservative/signal-flow semantics. The major difference is that VHDL-AMS is equation oriented, while Verilog-AMS is circuit oriented.

For the moment, the limitations are not due to the languages, but to operating simulation environments that do not implement all the language statements, or badly support them. Nevertheless, it is now possible to imagine the future of SoC and MOEMS design as the incremental interconnexion of composite pluri-disciplinary objects.

REFERENCES

[1] E. Christen, K. Bakalar, "VHDL-AMS-a hardware description language for analog and mixed-signal applications," IEEE Trans. on Circuits and Systems, part II, Vol. 46 Issue: 10, pp. 1263-1272, Oct. 1999.

[2] 1076.1-1999 IEEE Standard VHDL Analog and Mixed-SignalExtensions [ISBN 0-7381-1640-8]

[3] http://www.eda.org/verilog-ams/

[4] D. Fitzpatrick, I. Miller. 'Analog Behavioral Modeling with the Verilog-A Language' , ISBN 0-7923-8044-4

[5] http://www.verilog-2001.com/

[6] M. Bucher, C. Lallement, C. Enz, F. Théodoloz, K. Krummenacher, "The EPFL-EKV MOSFET Model Equations for simulation, version 2.6," http://legwww.epfl.ch/ekv/

[7] W. Uhring, Y. Herve, F. Pêcheux, "Model of an instrumented optoelectronic transmission system in HDL-A and VHDL-AMS," Proceedings of SPIE Volume: 3893, pp. 137-146, Design, Characterization, and Packaging for MEMS and Microelectronics, Editor(s): Bernard Courtois, Serge N. Demidenko. Published: 10/1999

[8] C. Lallement, R. Bouchakour and T. Maurel, "One-dimensional Analytical Modeling of the VDMOS Transistor Taking Into Account the Thermoelectric Interactions," IEEE Transactions on Circuits and Systems, Part. I, Vol. 44, N° 2, pp. 103-111, Feb. 1997.

[9] C. Lallement, F. Pêcheux, Y. Hervé, "VHDL-AMS design of a MOST model including deep submicron and thermal-electronic effects", IEEE Workshop BMAS 2001, pp.91-96, Santa Rosa, USA, 2001.

[10] C. Lallement, F. Pêcheux, Y. Hervé, "A VHDL-AMS Case Study : The Incremental Design of an efficient of a 3rd generation MOS Model of a Deep Submicron Transistor," SOC Design Methodologies, Ed. M. Robert, B. Rouzeyre, C. Piguet, M.L. Flottes, Kluwer Academic Publishers, ISBN 1-4020-7148-5, pp. 349-360, July 2002.

[11] F. Pêcheux, C. Lallement, "VHDL-AMS and Verilog-AMS as Competitive Solutions for the High-Level Description of Thermoelectrical Interactions in Opto-Electronic Interconnection Schemes", Proceedings FDL02, Marseille, France, Sept. 24-27, 2002.

Chapter 5

VHDL-AMS IN MEMS DESIGN FLOW

Joachim Haase, Jens Bastian, and Sven Reitz
Fraunhofer-Institute Integrated Circuits - Branch Lab Design Automation Dresden

Abstract: Behavioral modeling languages can be used in different steps of top-down design and bottom-up verification of MEMS design. The available language facilities of VHDL-AMS and the requirements of the applied methods in this process are confronted. Especially the application of Kirchhoffian networks to model 3D movements is taken into account. The decisions that should be done at the beginning of the modeling process are discussed. This is especially important if models from different sources shall be combined later on. Experiences using available simulation engines are presented. A micro mechanical accelerometer and an electrostatic beam actuator are investigated. Therefore a set of basic elements for MEMS simulation was created.

Key words: VHDL-AMS, MEMS, Kirchhoffian networks

1. INTRODUCTION

MEMS technology requires CAD tools for support. Modeling and simulation play an important rule in this environment [1], [2]. Compared to the design of electronic systems the development of MEMS design tools is at the beginning. But a lot of well-established ideas from EDA can be applied to MEMS design. Thus, a top-down methodology can be used to handle complex designs [3]. Behavioral modeling facilities allow to specify components of micro electromechanical systems and to simulate the specified systems. After the design or decision concerning the re-use of existing subsystems a validation of the system behavior should be done. This can also be carried out by simulation. Therefore behavioral models of the components with calibrated parameters based on the realized subsystems

E. Villar and J. Mermet (eds.), System Specification and Design Languages, 51–60.
© 2003 *Kluwer Academic Publishers.*

must be available for the simulation. This step is usually called bottom-up verification in the electronic design flow [4].

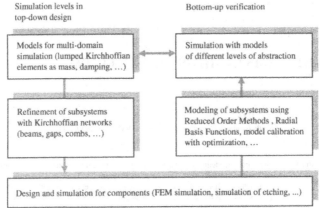

Figure 1. Top-down design and bottom-up verification in MEMS design flow

Advantages of the application of a unified behavioral description language in a MEMS design flow shown in Figure 1 are among other things the possibilities to combine models of different levels of abstraction and from different sources. One language that covers these requirements and the applied methods in the design flow is VHDL-AMS [5]. VHDL-AMS can be used to build complex analog and mixed-signal models. Differential equations, algebraic constraints and logic controls can be combined. With respect to the status of the standardization of the language and the scope of the language that is covered by commercially available tools, different implementation approaches are considered in the following.

2.   DEPLOYED MODELING METHODS

Modeling with Kirchhoffian networks is well established in MEMS simulation [6], [7]. The fundamental idea consists in the combination of subsystem models to describe a more complex behavior. The connection points carry across and flow quantities. The across quantities describe translational displacements in a global coordinate system, rotations about global axes, and electrical voltages. The flow quantities characterize forces in the direction of the coordinate axes, torques about axes, and currents resp.

Figure 2. Global coordinate system

The sums of the mechanical flow quantities at a connection point have to be zero for each axis of the coordinate system (see Figure 2). The connectivity of the components is described by a netlist. The simulation algorithm is responsible for the fulfillment of the Kirchhoff Laws. Refinement of the specification of mechanical subsystems should take into account the movement described in a global coordinate system. In the simpler case movements in x and y directions and rotations about the z axis are considered. That means only movements in the xy-plane are allowed in the simpler case. In the general case movements in all directions and about all axes must be taken into account [6], [8].

From the VHDL-AMS language point of view appropriate NATURE and TERMINAL declarations have to be applied to take into account the special MEMS requirements. Thus, a connection point of a MEMS component has to be characterized by a collection of different terminal declarations.

Table 1. Possibilities for terminal declarations

Displacement in xy-plane and rotation about z axis (3 mechanical degrees of freedom)	Displacements in all planes and about all axes (6 mechanical degrees of freedom)
terminal tx : kinematic; **terminal** ty : kinematic; **terminal** rz : rotational; **terminal** e : electrical;	**terminal** t : kinematic_vector (1 **to** 3); **terminal** r : rotational_vector(1 **to** 3); **terminal** e : electrical;

Different possibilities to define connection points are shown in Table 1. The declaration in the first column does not require the implementation of multi-dimensional terminals in a VHDL-AMS simulation engine. The declaration in the second column would allow an easy change from a lower dimensional description to a higher dimensional one by replacing the ranges of the array declarations. Unfortunately, it is not possible to combine the scalar and array natures that characterize a "MEMS point" in a composite record nature. Nature record fields have to be of the same type (see [9], section 3.5.2.2). Thus the current VHDL-AMS standard in principle only allows to handle "MEMS" points in ways similar to those suggested in Table 1. Furthermore, if the system of equations is established using SI units the solutions vary across several orders of magnitude [10]. This fact has to be taken into consideration during the solution of the simulation problem. One way would be to define appropriate natures and TOLERANCE groups for this purpose. At the moment, such declarations are not part of the draft packages of the standard package working group [11]. Another way is to equilibrate across and through quantities for the mechanical part using scaling factors.

The following rules were applied for the investigations:

- A multi-dimensional connection "MEMS point" is characterized in accordance with Table 1. The corresponding declaration of the left column can be implemented in most current VHDL-AMS simulation engines. Simple lumped elements as mass, spring, and damping elements only take into consideration displacements in one direction or rotations about one axis. NATURE, library and package identifiers are used in accordance with ADVance MS [12].

- If a subsystem is a pure mechanical one (as for instance a beam) only the mechanical part of the "MEMS point" description is used. Also in other cases only the necessary parts of the description are used.

- Forces and displacements of the model descriptions can be transformed to the SI system by multiplication with a factor scale_pos. In the case of angles and torques the values have to be multiplied by scale_ang.

- External forces are modeled by a through source that is directed from the concerning node to the corresponding reference node.

2.1   Analytical Models for Top-Down Design

The system of differential equations

$$M \cdot \ddot{x} + D \cdot \dot{x} + K \cdot x = -f \qquad (1)$$

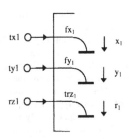

Figure 3. Model structure

describes the behavior of the (linear) mechanical part of the subsystems. M, D, and K are the analytically given real valued $N \times N$ mass, damping, and stiffness matrices. x is e. g. the vector of displacements and angles of rotation in a global coordinate system. f is the contribution vector of forces and torques at the connection points of the subsystem that are responsible for the across quantities x. x and f are quantities $x, f : [0, Tend] \rightarrow R^N$.

Figure 3 shows a part of the internal structure of a model described by (1). The (scalar mechanical) terminals are connected by branches with the associated reference nodes. x_1, y_1, and r_1 are components of a vector x. The through quantities fx_1, fy_1, and trz_1 are components of a vector f. A good choice of fundamental MEMS elements was proposed in [7] and following papers. The advantage of a VHDL-AMS solution is the possibility to combine all these models with other models and create user-defined models in an easy way.

Example: The following listing shows the interface of a mechanical beam (see e. g. [7]) that can be moved in the xy-plane. General technology data are stored as deferred constants in the package mems_technology of the resource library mems. Default values of the generic parameters are taken from this package. The package mems_technology, especially the package body where the values are assigned to the constants, has to be replaced in order to change from one technology to another one.

```
library disciplines, mems;
use disciplines.kinematic_system.all;                 o——| tx1    tx2 |——o
use disciplines.rotational_system.all;
use mems.mems_technology.all;                          o——| ty1    ty2 |——o

                                                      o——| rz1    rz2 |——o
entity beam2d is
    generic
        (l : real;                         -- beam length in meters
         w : real;                         -- beam width  in meters
         h : real := mems.mems_technology.h; -- beam thickness in m
         density  : real := mems.mems_technology.density;
         -- ...
         oz : real := 0.0);      -- initial rotation about beam's z-axis
    port (terminal tx1, ty1 : kinematic;
          terminal rz1       : rotational;
          terminal tx2, ty2 : kinematic;
          terminal rz2       : rotational);
end entity beam2d;
```

Listing 1. Entity declaration of a mechanical beam

2.2 Reduced Order Models for Bottom-Up Verification

In many cases the real design of the non electrical components is finally carried out with the help of simulation programs that can solve systems of partial differential equations such as the FEM tool ANSYS. For bottom-up verification order reduction methods allow to derive descriptions from linear PDEs that can be used to establish behavioral models. In contrast to (1), the matrices cannot be expressed by analytical formulas that depend on generic parameters. They are built up by fixed values. The equations of a reduced order model are given by

$$\widetilde{M} \cdot \ddot{\widetilde{x}} + \widetilde{D} \cdot \dot{\widetilde{x}} + \widetilde{K} \cdot \widetilde{x} = -\widetilde{B} \cdot f \quad \text{and} \quad x = \widetilde{B}_a \cdot \widetilde{x} \qquad (2)$$

$\widetilde{x}$ is a vector of auxiliary quantities of the reduced order model. $\widetilde{M}$, $\widetilde{D}$, and $\widetilde{K}$ are the reduced mass, damping, and stiffness matrices. The matrices of the reduced order model can be derived using projection methods from the system of ordinary differential equations that results from the semi-discretization of the PDEs [13]. $\tilde{B}$ is the incidence matrix of connection points with applied forces, torques etc., f. $\tilde{B}_a$ is the incidence matrix to observe across quantities x at the connection points. Usually $\tilde{B}_a$ equals $\tilde{B}$. In a VHDL-AMS model, $\tilde{x}$ can be declared as a vector of free quantities. x and f can be declared similar to Figure 3. A natural way would be to read the values of $\tilde{M}$, $\tilde{D}$, and $\tilde{K}$ from a file. That means the I/O-facilities of VHDL-AMS are very helpful especially during the bottom-up verification.

Example: The reduced order model only takes into consideration the translational movement in one direction. The reduced matrices are read from a file with the user-defined function `read` into arrays `mt`, `dt`, `kt` and `bt` of type `real_vector`. The free quantities `xt`, `dxt`, `d2xt` correspond to $\tilde{x}$ and its derivatives. The local function `reduced_model` applies user-defined overloaded operators from the package `operator` of library `mems` to implement the matrix vector multiplications of (2). The model was successfully applied in time domain simulation.

```
library disciplines, mems;
use disciplines.kinematic_system.all;
use mems.operator.all;

entity sensor is
   generic (n          : integer := 6;
            file_name  : string  := "tilde.dat");
   port    (terminal t1 : kinematic);
end entity sensor;

architecture reduced of sensor is
   constant mt  : real_vector (1 to n*n)
                    := mems.encapsulation.read ("m"&file_name);
   -- ... (read dt, kt, bt)
   quantity xt, dxt, d2xt        : real_vector (1 to n);
   quantity x across f through t1;
   constant lhs : real_vector (1 to n) := (others => 0.0);

   function reduced_model (n : integer;
          mt, dt, kt, d2xt, dxt, xt, bt : real_vector; f : real)
   return real_vector is
      variable i1, i2 : integer; variable r  : real_vector (1 to n);
   begin
      for i in 1 to n loop
          i1   := 1 + (i-1)*n; i2    := i*n;
          r(i) := mt(i1 to i2) * d2xt + dt(i1 to i2) * dxt
                  + kt(i1 to i2) * xt  + bt(i)*f;
      end loop;
      return r;
```

```
   end function reduced_model;

begin
   dxt  == xt'dot; d2xt == dxt'dot;
   lhs  == reduced_model (n, mt, dt, kt, d2xt, dxt, xt, bt, f);
   x    == bt*xt;
end architecture reduced;
```

Listing 2. Example of a reduced order model

2.3 Language Requirements

Besides the discussed and used language statements also other facilities of VHDL-AMS are helpful for MEMS modeling. For instance micro electromechanical systems often consist of regular structures. The GENERATE statement of VHDL-AMS that can be applied on concurrent and simultaneous statements may help to describe such structures (see example in the next section). Comprising we can say that the following VHDL-AMS language constructs are helpful to support modeling and simulation in the MEMS design flow (especially if Kirchhoffian networks are used)

- Concept of entity and architectures to describe a system on different levels of abstraction
- Declaration of packages with typical technology data
- Declaration of functions that are widely used in MEMS models
- Overloading of at least addition and multiplication for real-valued one- and multidimensional arrays and their application in simultaneous statements
- Usage of file I/O operations
- Description of regular structures with the GENERATE statement

It depends on the target simulators which of these facilities can be used. Models can be exchanged if the corresponding language constructs are available in the target simulation engines. Furthermore some general developments would support MEMS modeling

- Declaration of mixed RECORD natures
- Declaration of the across and through subtypes with special tolerance aspects for MEMS applications

Compared to other solutions the great advantage of VHDL-AMS is the availability of the usual simulation algorithms (DC, AC and transient) in a dedicated simulation engine. Typical MEMS question as e.g. the determination of resonant frequencies and values of pull-in voltages can be answered by the formulation of adequate simulation problems.

3.   EXAMPLE AND EXPERIENCES

On this basis some elementary Kirchhoffian elements (similar to those used in SUGAR) were implemented using VHDL-AMS. Beam models describe linear mechanical beams that are in-plane and can rotate out-of-plane. Gap models describe electrostatic gaps, which consist of two mechanical beams that are the movable electrodes. Models of external forces, anchors, electrostatic comb, mass, spring, and damping are also available. Besides these elements all other facilities of VHDL-AMS can be used. With these elements some subsystem specifications were simulated in

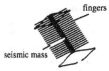

Figure 4. Movable parts of sensor

correspondence to Figure 1. Figure 4 represents a micro mechanical accelerometer. The picture was created with the help of SUGAR [10]. The sensor consists of a movable beam (seismic mass), suspended by two spring tethers on either end. Movable fingers are attached to the mass. The fingers establish together with fixed plates (not represented in Figure 4) capacitances that are evaluated by an electronic circuit. If the seismic mass is moved by an external force, the capacitances depend on this force. The structure is highly regular. The model consists of mechanical beams of different dimensions. The electrostatic forces are modeled by comb

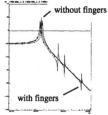

Figure 5. Transfer characteristics

models. Figure 4 shows the displacement of the sensing element at the first resonant frequency. The transfer characteristics are simulated for different modeling approaches using AC analysis. The transfer characteristics describe the displacement of the seismic mass for different frequencies.

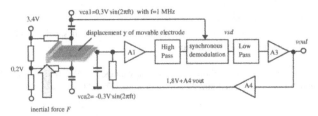

Figure 6. Sensing element and electronic circuitry [14]

Figure 7. Simulation results

Similar micro mechanical devices are used in force-balanced accelerometers like ADXL series from Analog Devices and Siemens. Figure 6 shows a schematic drawing of the sensing element and the electronic circuitry of such an accelerometer. The inertial force *F* causes the movable fingers to displace. This displacement y causes a voltage in phase with the carrier vca1 at the input terminal of the buffer amplifier A1. The signal is amplified. The output of the high pass filter *v01* is fed to the synchronous demodulator. The output *vsd* is filtered by a low pass filter and amplified. The output of the amplifier A3 is the output *vout* of the circuit. The output signal is also fed back to the seismic mass. In more detail the circuit is described in [14].

The basic functionality can easy be simulated using VHDL-AMS. For test purposes behavioral models of the electronic subsystems and a model of the sensing element using basic MEMS primitives (see section 2.1) were used. The advantage of this approach is an easy combination of these MEMS primitives with other user-defined models.

4. CONCLUSION

VHDL-AMS covers the requirements of different methods applied in the MEMS design process. The IEEE standard [9] offers very good possibilities that support these objectives. With respect to the implementation of the standard in available simulation engines some of the facilities cannot be applied at the moment to obtain models that can be exchanged between different simulators. This has to be taken into consideration during the creation of models. Nevertheless practical examples can be modeled and simulated using the available language constructs.

Parts of this contribution were founded by German Bundesministerium für Bildung, Wissenschaft und Technologie (Project EKOSAS "Entwurf komplexer Sensor-Aktor-Systeme", Label 16 SV 1161).

REFERENCES

[1]   Senturia, S. D.: CAD Challenges for Microsensors, Microactuators, and Microsystems. Proceedings of the IEEE 86 (1998) 8, pp. 1611-1625.

[2]   Fedder, G. K.; Jing, Q.: A Hierachical Circuit-Level Design Methodology for Micromechanical Systems. IEEE Trans. on CAS-II 46(1999)10, pp. 1309-1315.

[3]   Fedder, G. K.: Top-Down Design of MEMS. Technical Proceedings of the International Conference on Modeling and Simulation of Microsystems (MSM 2000), San Diego, CA, March 27-29, 2000.

[4]   Oudinot, J.; Vaganay, C.; Robbe, M.; Radja, P.: Mixed-Signal ASIC top-down and bottom-up design methodology using VHDL-AMS. Proc. XV Conference on Design of Circuits and Integrated Systems, Montpellier, November 21-24, 2000.

[5]   Dewey, A.; Dussault, H. et al: Energy-Based Characterization of Micromechanical Systems (MEMS) and Component Modeling Using VHDL-AMS. Proc. Modeling and Simulation of Microsystems, Semiconductors, Sensors, and Actuators (MSM 99), San Juan, Puerto Rico, April 19-21, 1999.

[6]   Neul, R.; Becker, U.; Lorenz, G.; Schwarz, P.; Haase, J.; Wünsche, S.: A modeling approach to include mechanical microsystem components into system simulation. Proc. Design, Automation and Test in Europe (DATE'98), Paris, February 1998, pp. 510-517.

[7]   Clark, J.V.; Zhou, N.; Pister, K.S.J.: MEMS Simulation using SUGAR v0.5. Tech. Digest, Solid-State and Actuator Workshop, Hilton Head Island SC, pp. 191-196, June 8-11, 1998.

[8]   Clark, J. V.; Bindel, D. et al: 3D MEMS Simulation Modeling Using Modified Nodal Analysis. Proc. of the Microscale Systems: Mechanics and Measurements Symposium, Orlando, June 8, 2000, pp. 68-75 (http://bsac.berkeley.edu/cadtools/sugar/sugar/).

[9]   IEEE Std 1076.1-1999 Standard VHDL Analog and Mixed-Signal Extensions. The Institute of Electrical and Electronic Engineers, 1999.

[10] Bai, Z.; Bindel, D. et al: New Numerical Techniques and Tools in SUGAR for 3D MEMS Simulation. Technical Proceedings of the International Conference on Modeling and Simulation of Microsystems (MSM 2001), Head Island, March 19-21, 2001 (http://bsac.berkeley.edu/cadtools/sugar/sugar/).

[11] Information of Standard Package IEEE Working Group (http://www.vhdl.org/analog/)

[12] ADVance   MS   simulation   engine   of   Mentor   Graphics,   Inc. (http://www.mentor.com/ams/adms.html)

[13] Sheehan, B. N.: ENOR: Model Order Reduction of RLC Circuits Using Nodal Equations for Efficient Factorization. Proc. 36th Design Automation Conference, 1999, pp. 17-21.

[14] Bao, M.-H.: Micro Mechanical Transducers. Pressure Sensors, Accelerometers, and Gyroscopes. Amsterdam: Elsevier Science, 2000.

Chapter 6

A NEW APPROACH TO MODEL GENERATION FOR NONLINEAR MIXED-SIGNAL CIRCUITS IN THE BEHAVIORAL AND FUNCTIONAL DOMAIN

Ralf Rosenberger and Sorin A. Huss
Darmstadt University of Technology; Department of Computer Science; Alexanderstr. 10, 64283 Darmstadt, Germany; rosenberger@iss.tu-darmstadt.de, huss@iss.tu-darmstadt.de

Abstract: *In the design process of mixed-signal circuits, the analog part is handled in many areas separately from the digital part. Models for analog blocks, as reported in literature, are available on different levels of abstraction. They are implemented in analog simulation languages, which in general do not support digital signals and thus need artificial interfaces in order to interact with the digital world. VHDL-AMS allows the combination of analog and digital models, but common approaches for mixed-signal modeling are in the functional domain only, where the analog signals are not physical any more. In this paper, a new approach to model nonlinear mixed-signal functional blocks is proposed. The outlined methodology combines analog and digital signals within one joint model, reflects the nonlinearity of complex analog modules and maintains at the same time the physical properties of electrical signals.*

Key words: Mixed-signal simulation, VHDL-AMS, Model generation

1.   INTRODUCTION

The in-depth analysis of a real system requires costly methods of the physical measurement technology. Furthermore, it assumes the existence of at least one prototype. In order to gain insights into the behavior, there are computer-based methods to simulate such a system. Computing the detailed

E. Villar and J. Mermet (eds.), System Specification and Design Languages, 61–73.
© 2003 *Kluwer Academic Publishers.*

behavior of functional blocks and of complete systems is of utmost interest both to a modeler and to a circuit designer.

Abstraction hierarchies and resulting properties of models are common topics right from the beginning of simulation methods. For analog components [14] and digital modules [6] a hierarchy of abstraction levels and transformations between different levels is widely accepted. Mixed-signal systems, however, need a special view on the interaction of continuous and discrete signals. Therefore, Figure 1 depicts a new abstraction hierarchy for such systems in order to address this problem accordingly [8][13]. In contrast to the functional domain, signals being present in behavioral models are subjected to conservation laws (i.e., the Kirchhoff rules are to be met).

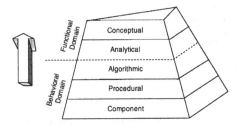

Figure 1. Levels of abstraction for mixed-signal circuits.

Conceptual Level. The topmost level of the functional model class may be present. It possibly exploits temporal logic and some flow calculus for functionality specification.

Analytical Level. In the functional model class, the analytical level is assumed to be the highest abstraction level of practical interest. It addresses a specification of functionality by means of analytical equations for continuous signals and of abstract processing for discrete signals. The fundamental concept on this level is causal modeling.

Algorithmic Level. In behavioral models both the control and the data flow have to be denoted at algorithmic level. Continuous behavior is refined into disjoint operation regions of the model, which are to be represented accordingly by sets of analytical equations (i.e., data flow). Selections on the explicit equations to be executed at a given point in time take place according to parameter values, which define the active operation domain (i.e., control flow). Algorithmic level models exploit in general concurrent execution concepts of at least the data flow.

Procedural Level. This level addresses refinement operations of an algorithmic model into modules. The resulting modules (i.e., procedures of the model code), represent either model entities of available building blocks or processing units in a similar way to a structured software code.

Component Level. The component level is aimed to a direct representation of structural netlists of circuits, which are composed from basic elements, such as transistors or logic gates. Their behavior has to be defined as part of the associated component library models.

An approach to model blocks in the functional and behavioral domain is advocated in the following, which is aimed to support nonlinear behavior in time domain operation. Figure 2 summarizes the internal structure of the model constructed for that purpose: The concept is to model the relation between the input and output stages exploits the acausal modeling paradigm for continuous signals combined with event processing algorithms. A dynamic consideration of the load situation at the output enables the conservation laws to be considered. Thus, the internal model operation at the analytical level is transferred into the behavioral domain according to Figure 1.

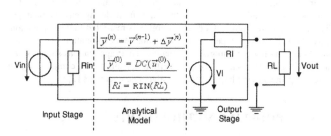

Figure 6-2. Generic structure of block model.

The remainder of this paper focuses on the key aspects of the modeling methodology, the underlying systems theoretic background and a detailed application example. Finally, a discussion of results and ideas for further work concludes the paper.

2. RELATED WORK

Common approaches and methods to generate models of mixed-signal circuits can be classified into three groups:

– Structural netlists at component level,
– Macro models at procedural level,
– Behavioral models at algorithmic level.

Structural netlists are the well-known SPICE netlist models for analog circuits. Models can be built in a hierarchical way (Bottom-Up design), yielding a direct mapping of the circuit topology to the model representation. Their main disadvantage is the demand of simulation resources. Mixed signal circuits cannot be represented at all due to a lack of an event-processing engine.

Models at procedural level are often provided by the circuit vendor. Numerical methods, such as [5] [4] are frequently of limited accuracy especially for strongly nonlinear blocks. Empirical models (e.g., [3] [9]) rely on analytical equations and subcircuits, which are mapped to either basic or compound elements. In general, these models suffer from an unpredictable accuracy and from technology dependence.

Algorithmic models are distinguished by the fact that the mapping of input to output signals is described completely by mathematical equations and event processing algorithms. The analog voltages and currents are abstracted into time and value continuous signals. An example of a commercially available toolset is the Antrim modeling environment [1]. These models are empirical, i.e., the usability is restricted to the application-cases, which the developer did into account. An exploitation of computer algebra methods is an important alternative. Their applications shown so far unveil, in general, a lack of run time efficiency due to the complex task of setting up and solving large systems of algebraic equations [2].

3. MODELING METHODOLOGY

This chapter details how the modeling methodology for functional blocks is realized. The steps that need to be performed by the circuit designer to generate an executable model of the functional block starting with a representation of the circuit at a low level of abstraction are illustrated in Figure 3. As inputs, the representation of the circuit in the HDL of the simulator in use, the characterization-plan and the requirements in terms of precision are needed. The result is the executable model of the functional block in a standard HDL - e.g. VHDL-AMS [12].

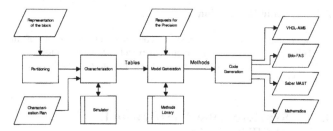

Figure 3. Model generation flow.

3.1   Partitioning

An appropriate partitioning of the circuit is important, especially when building models for large circuits. The resources required for the characterization (the size of the function-block correlates with the time required for the characterization) and for the execution depend on the chosen partitioning. The area of application of the modeling approach can be extended by subdividing a circuit that cannot be modeled as a whole into several partitions that fulfill the requirements.

3.2   Characterization

The characterization takes most of the computing time within the modeling flow. The network simulator is running under "remote" control, i.e., all simulations necessary to execute the characterization plan on the representation of the circuit run automatically and the circuit characteristics required to setup the model are extracted accordingly. The results of the characterization are tables [7].

3.3   Model Generation

The tables that result from the characterization run are then converted to mathematical equations used in the function-block-model by means of the methods-library considering the precision goals [11]. One function can be modeled with different resolution values, e.g., when executing a model at analytical level, an efficient model with a larger maximum error can be used, while verification at algorithmic level requires a model with higher precision.

3.4 Code Generation

The mathematical model must be converted into the HDL of the simulation environment in order to provide executable representations.

The result is the function block model produced according to the proposed methodology.

4. SYSTEMS-THEORETIC MODEL ARCHITECTURE

Systems theory aims to provide generally applicable problem solving methods for different disciplines in science. The specification formalisms are well known (e.g., [15]). The simulation of mixed signal circuits requires models that contain continuous as well as discrete parts. The necessary systems theoretical foundations for this application are given in [10].

The block models are defined starting from a combined type of formalism, the Differential Equation Specified System with Discrete Events (DESS&DEVS) [10]:

$$BlockModel = \{X, Y, S, \delta_{\text{int}}, \lambda, ta, f\}$$

Set of inputs, outputs and internal states

$$X = \{x_i \mid x_i \in \mathbb{R}, i = 1...n\}$$
$$Y = \{y_i \mid y_i \in \mathbb{R}, i = 1...m\}$$
$$S = \{s_i \mid s_i \in \mathbb{R}, i = 1...k\}$$

The internal state variables are intended to capture the functionality of the block. Currents and voltage differences are examples for such state variables in electrical circuits. The set of outputs of the block can be calculated at any time step by means of internal state and input values.

State transition function for internal events

$$\delta_{\text{int}}(S) = \{m_i(S) \mid m_i(S) \in Methods_{Block}, i = 1...l\}$$

The internal state transition function is evaluated when a scheduled event is processed, it updates the internal state variables. The internal state transition function is represented by a set of methods, which are generated during the model generation.

Output function $\lambda(S, X) :\to Y$

The output function is a recursive function. Its parameters are input and internal states.

Time advance function $ta :\to ta_{ext} \cup ta_{int}$

The time function schedules both the external and internal events.

Rate of change function $f :\to \dfrac{\delta s}{\delta t}$

The rate of change function calculates the continuous change of one or more internal state variables.

The systems-theoretic model as defined above gives the signature for the elements included in equation form. This formal specification consists of intrinsic functions m_i, which are defined in dependence on the input and state variables. These intrinsic functions are related to the state transition function $\delta()$ and rate of change function f. In each integration step the values of m_i correlating to the values of the input and state variables must be calculated. A simulator run of the reference circuit representation with the inputs set accordingly is to be performed. Then, the values of m_i can be extracted from the resulting tables. The simulations can be run in the DC- or Transient domain. In order to further speed-up execution and reduce the amount of storage needed the tables are replaced by mathematical models, the so-called Methods. A method is defined as a set of mappings of data segments to approximating mathematical functions, according completeness, uniqueness, C_0-smoothness and accuracy requirements. Well-known error norms l_1, l_2 and l_∞ and different distance metrics can be applied to control the accuracy level of the approximation.

The last step in generating the executable model is the conversion of the methods to HDL-Code. Now, a ready to use representation of the block model in the simulation environment is available.

5.   APPLICATION EXAMPLE

The outlined methodology as detailed in [12] is demonstrated for an AD Converter - a complex mixed-signal circuit with a single analog input and each one digital output per bit. The simulation of the circuit requires high resources in terms of computation time: A simulation run for a 6 bit AD Converter on an UltraSparc workstation with 2 processors (300MHz) takes approximately 11 min when exploiting a SPICE-like simulator. The 6 bit AD converter consists of 5 converter stages and one comparator, the architecture of a converter stage is detailed in Figure 4. As representations at a lower level of abstraction, analog macro-models delivered by the respective manufacturer were used. The reference-model is denoted as procedural model according to the definitions in Chapter 1.

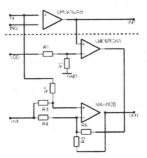

Figure 4. Converter stage of AD converter.

5.1   Characterization and Model Generation

The analog output signal of the converter stage is not smooth: When the bit output changes its value, a sudden shift of the analog signal takes place. This shift cannot be modeled directly. Thus, the converter stage is partitioned into two parts: The comparator, responsible for the shift, is removed from the subcircuit. The shaded line in Figure 4 shows the partitioning. This subcircuit has single analog and digital inputs and a single analog output only.

The tables resulting from the characterization are converted in mathematical models, the methods. These are converted into VHDL-AMS code and copied into the feasible model. The A/D converter model includes the methods DC output and slewrate twice, to distinguish between bit input on and off to reflect the hysteresis curve correctly.

5.2   VHDL-AMS code of the block model

The entity generic_CONVSTAGE has the implementation of the block model for the subcircuit in VHDL-AMS . The following code fragment shows the implementation of one methods and parts of the main simulation loop.

```
architecture behavioral of generic_CONVSTAGE is
-- Begin ******* Method slewrate0 ********
function slewrate0(Vdiff: real) return  real is
variable V_diff, SR: real;
begin
if V_diff <= 2.96e-6 then
   SR := 0.0;
elsif  V_diff <= 1.419e-1 then
   SR := -4.4832957e3 - 1.1722595e6 * V_diff
+1.7897751e9 * V_diff ** 4.0e0 - 5.8280225e8 * V_diff **
3.0e0 + 6.0415514e7 * V_diff ** 2.0e0;
elsif  V_diff <= 3.9859e-1 then
   SR := 3.1599691e4 + 6.3580799e5 * V_diff -
7.7975484e5 * V_diff ** 2.0e0;
elsif  V_diff <= 1.61786e0 then
   SR := 1.5973821e5 + 3.5260018e3 * V_diff;
else SR := 165442.786;
end if;
return SR;
end;
-- End ******* Method slewrate0  ********

begin
   -- calculation of effective Output Difference
   VI == VOUTDC1( U_in ) - U_out;
   -- calculation of the output voltage
   if (now<TD and Bit) use
      U_out'dot == slewrate0( VI'Delayed( TD ));
   end use;
   -- discontinouities
   break when U_out'above(V_out_max);
   break when not(U_out'above(V_out_min));
end behavioral;
```

The full model uses this converter stage as subcircuit to construct the 6-Bit ADC. The ADC is included in the testbench to compare the block model to different modeling approaches.

5.3   Simulation Results for the 6 Bit ADC

The 6 bit A/D converter consists of 5 converter stages. The analog output of the first stage is connected to the analog input of the second stage and so on. Thus, errors from the first stage are propagated to the last stage, the errors from the different stages are overlaid and add up. The slope of the analog output must be modeled with high accuracy to guarantee the correct switching times of the following comparators, resulting in possible errors in phase. Figure 5 shows the simulation result in the time domain of the full converter stage with a sinus input. s for the 6 Bit ADC

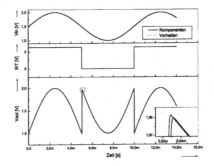

Figure 5. Comparison of the simulation results for the converter stage.

Here, the difference between the block model and widely used models is emphasized: Since slope and delay are used within the model, the converters of the following stages switch correctly, the output signal is maintained without an error in phase.

To validate the correct behavior of the full converter, the block model is compared to the reference circuit. The testbench is stimulated with a sine-waveform. The comparison of the results in the time domain is shown in Figure 6 . The digital exits carries the correct signals, which implies the analog exits of the converter stage have an error in the rising signal of less than 1%. The precision can be improved further through accordingly higher expenses in the characterization part.

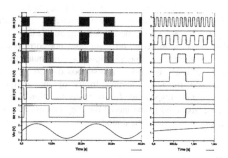

Figure 6. Simulation result for the 6 bit ADC.

Table shows a decrease of the simulation-time in the order of a magnitude for the converter. This reduces the typical simulation time from hours into the range of minutes.

Table 6-1. Comparison of simulation times.

	Speedup
Converter Stage	3,6
6-Bit A/D-Converter	12

5.4   Comparison of Modeling Accuracy

Three models of the A/D Converter which were developed on different levels of abstraction are compared to each other:
- **Procedural level**: Reference Model used for characterization.
- **Algorithmic level**: Model generated with the proposed approach.
- **Analytical level**: Model with a analytical (mathematical) relationship between input and output without reference to the conservation laws.

In the large-signal behavior, no differences appear between the models. Further tests take place in the time domain. Looking at the LSB of the output, changes at this digital signal of the analytical model occur at significant different times. The reason is the additive overlay of small errors in the switch times: The slope of the analog output of the subcircuit is limited. Since the step in the analytical-model is ideal (unlimited slope), the comparator of the next converter stage switches too early. This mistake propagates up to the fifth stage and increases about one order of magnitude in each stage.

In another example, a 10 kHz sine waveform is applied to the converter. This frequency is too large for the reference circuit: The output cannot follow the input. The analytical model delivers an ideal signal, which is mathematically correct admittedly, however does not correspond to reality. The results of the algorithmic model allows a clear qualitative statement however: The circuit does not work as it is supposed to do.

This comparison illustrates certain limits of the analytical model. Physical effects, that are responsible for the faulty-function of the procedural model, were not considered in the model building process and, accordingly, are not reflected in generic analytical models.

6.   CONCLUSIONS

The proposed approach to modeling of mixed-signal functional blocks results in a considerable increase of simulation speed while preserving to a large extent the shape of signal waveforms. Its main advantages however are the reusability of the models in different circuits with arbitrary load conditions and the automatic model generation without empirical knowledge of the model designer. No details of the concrete implementation of the circuit can be extracted from the model. The model can be scaled for simulations in the behavioral and functional domain. The interdisciplinary systems theory background forms a basis for an easy mapping to different disciplines. Digital signals do not need interface applications: The mixed-mode capabilities of VHDL-AMS allow an easy adoption of the approach presented.

The actual implementation needs further improvements. Applications of the proposed approach are limited to circuits that show a step-response at the output. Especially circuits with oscillating outputs are not supported. An integration in a visual programming environment should free a human modeler from programming details when generating the models. Small-signal effects need to be further examined. Applications from different engineering disciplines must be worked out in order to demonstrate that the approach is interdisciplinary valid.

REFERENCES

[1] Antrim Mixed-Signal Characterization and Library Generation, http://www.antrim.com.

[2] C. Borchers, Automatische Generierung symbolischer Verhaltensmodelle für nichtlineare Analogschaltungen, 4. GMM/ITG Diskussionstagung Entwicklung von Analogschaltungen mit CAE-Methoden, pages 93--98, 1996.

[3] R. Boyle, M. Cohn, and O. Pederson, Macromodeling of Integrated Circuit Operational Amplifiers, IEEE Solid-State Circ., SC-9(6):353--64, 1974.

[4] G. Casinovi and A. Sangiovanni-Vincentelli, A Macromodeling Algorithm for Analog Circuits, IEEE Trans. on CAD, 10(2):150--60, 1991.

[5] L.O. Chua and P.M. Lin, Computer-Aided Analysis of Electronic Circuits: Algorithms and Computational Techniques, Prentice Hall, 1975.

[6] D. Gajski, N. Dutt, A. Wu, and S. Lin, High-level synthesis: Introduction to Chip and System Design, Kluwer Academic Publishers, 1992.

[7] M. Goedecke, A Visual Simulation Environment for efficient characterization of analog circuit behavior, Dissertation, Technische Universität Darmstadt, FB Informatik, Fachgebiet Integrierte Schaltungen und Systeme, 2001.

[8] S. A. Huss, A Guide to Generating Accurate Behavioral Models in VHDL-AMS, In Model Engineering in Mixed-Signal Circuit Design, Kluwer Academic Press, 2001.

[9] H.T. Mammen and W. Thronicke, Object-oriented Macromodelling of Analog Devices, In Proc. of the Internat. Conf. on Concurrent Engineering and Electronic Design Automation, pages 331--36, 1994.

[10] H. Praehofer, System Theoretic Foundations of Combined Discrete-Continous System Simulation, Phd thesis, Johannes Kepler Universitaet Linz, Department of Systems Theory, 1991.

[11] R. Rosenberger, Entwurf und Implementierung einer Methodenbibliothek zur Verhaltensmodellierung analoger Schaltungen, Diplomarbeit, Technische Hochschule Darmstadt, FB Informatik, Fachgebiet Integrierte Schaltungen und Systeme, 1994.

[12] R. Rosenberger, Zur Generierung von Verhaltensmodellen für gemischt analog/digitale Schaltungen auf Basis der Theorie dynamischer Systeme, Dissertation, Technische Universität Darmstadt, FB Informatik, Fachgebiet Integrierte Schaltungen und Systeme, 2001.

[13] S. A. Huss, S. Klupsch and R. Rosenberger, Modellierung gemischt analog/digitaler Schaltungen mit VHDL-AMS, In W. G. W. John, H. Luft, editor, 8. GMM Workshop Methoden und Werkzeuge zum Entwurf von Mikrosystemen, page Nachtrag. FhG IZM Advanced System Engineering, Paderborn, 1999.

[14] V. Moser, H. P. Amann and F. Pellandini, Behavioral Modeling of analogue Systems with absynth, In O.L. A. Vachoux, J.-M. Berge and J .Rouillard, editors, Analog and Mixed-Signal Hardware Description Languages, Kluwer Academic Press, 1997.

[15] B. P. Zeigler, Multifacetted Modelling and Discrete Event Simulation, Academic Press, 1984.

Chapter 7

MULTI-LEVEL ANALOG/MIXED-SIGNAL IP SPECIFICATION FOR PLATFORM-BASED DESIGN

Natividad Martínez Madrid[*], Antonio Acosta Fernández[**], Felipe Ruiz Moreno[**], Ralf Seepold[*]

[*] *Universidad Carlos III de Madrid, Dep. de Ingenieria Telematica*
Av. Universidad, 30, E-28911 Leganes (Madrid), Spain

[**] *Forschungszentrum Informatik (FZI) an der Universität Karlsruhe,*
Dept. Microelectronic System Design (SiM)
Haid-und-Neu-Str. 10-14, 76131 Karlsruhe, Germany

1. INTRODUCTION

The design of microelectronic systems is facing huge challenges in the last years. Some of these challenges are due to the steadily increasing complexity of the systems. According to Moore's Law, which has proven to be true so far, the integration capabilities on silicon will double every 18 months. Meanwhile, the capabilities allow the design of so-called System-on-Chip (SoC). Typically, SoCs include microprocessor cores, embedded software, DSPs, memories, peripheral controllers and mixed-signal parts.

The current and future trend in IP-based SoC design is based on integration platforms [4]. According to the predictions of the Semiconductor Industry Association, 70% of all integrated circuits will contain analog components in five years, compared to 25% nowadays. This contribution will analyze the requirements and constraints imposed by platform-based SoC design for analog and mixed-signal (AMS) IPs. In order to integrate AMS IP inside the complete design flow, several levels of abstraction need

E. Villar and J. Mermet (eds.), System Specification and Design Languages, 75–84.
© 2003 *Kluwer Academic Publishers.*

to be incorporated, and therefore, more views than only the current hard IP (layout) view are needed.

Two important roadmaps address the area of AMS reuse. The first one is the International Technology Roadmap for Semiconductors (ITRS) [16], which concentrates on technology issues and, in the 2001 edition, recognizes the predominant impact of design. In particular, design reuse is considered to be one of the system complexity challenges. According to the roadmap, the future key area in design reuse will be hierarchical design, heterogeneous SoC modeling and integration, with special emphasis on the consideration of mixed-signal systems. The second roadmap is the MEDEA+ EDA Roadmap [21]. Specially two chapters, the first dealing with IP reuse and the other with analogue design, highlight the relevance of AMS reuse for the European industry.

Several literature references exist in the area of design reuse [32, 33, 34], and some with particular focus on AMS reuse [20, 30]. This topic has also gained interest for the industry, which is leading relevant research and development activities in the area of AMS reuse [35]. The industrial relevance comes hand in hand with the higher availability of analog synthesis tools. These tools range from general, highly interactive synthesis tools from specifications down to layout generation [1, 5, 11, 22, 23, 29], through dedicated generation and layout migration tools [2, 3, 19].

Furthermore, standardization is crucial for AMS IP at all the different levels of abstraction. On the one hand, to allow a smooth integration of IP developed and acquired through different sources; and on the other hand, to allow a common understanding among the different actors and to increase the confidence of the user.

The main part of the paper will discuss industrial research activities that are under development. The first one, presented in Section 2, addresses the area of multi-level AMS IP characterization. Based on a market study, IP have been classified in a functional taxonomy, and a hierarchical set of attributes has been defined for each class in the taxonomy. This work is part of a proposal for taxonomy of the different levels of abstraction in AMS design. A prototype implementation of this work will be incorporated in the Reuse Management System (RMS) [24].

The second activity in the area of AMS IP reuse, presented in Section 3, focuses on system-level mixed-signal specification for reuse. There are currently many specification formalisms for mixed-signal systems at system-level covering different models of computation. In this approach, the main requirement for the selected language is to provide easy integration in SoC. In this sense, SystemC has been selected for implementation. The contribution in this area is to take into account reuse requirements for heterogeneous systems design while defining extensions to the language. These requirements will include a synthesizability analysis to connect to the lower levels of abstraction. The results of this work will be also linked to the RMS. The paper will conclude with a summary and a short discussion of future issues.

2. AMS IP CHARACTERIZATION

In the AMS domain, although not yet comparable to digital, several companies offer already AMS IP. A fundamental issue not yet solved is to manage the definition of adequate levels of abstraction for analog and mixed-signal behavioral description to link with digital top-down design.

Furthermore, formalized characterization of different analog and mixed-signal cores is required. The results of the AMS IP characterization process provide a view of the capabilities and ranges of performance of the design. A characterization takes into account all the different classes and subclasses of AMS IPs and a hierarchical set of attributes. The attributes will be typed. Numerical will have a unit, range or average. And the characterization will take care of the different levels of abstraction and SoC integration requirements.

2.1 Platform-based AMS methodology

The proposed platform-based mixed-signal design reuse methodology [12] is depicted in Figure 1. Following a top-down design, the requirements for the new system will guide the platform selection. Each platform has a particular portfolio associated, including AMS IPs, also called Virtual Components (VC). Before incorporating cores into a portfolio, a quality check will prove compliance to standards (VSIA) and to coding guidelines.

Each of the components will have a description at different levels of abstraction taking into account requirements for the integration in SoC. A formalized characterization of these models will allow automatic search and the comparison between similar cores a fast selection of the right core.

IPs will have also a behavioral description in order to allow a fast selection of cores and parameters. These high-level models will describe different non-ideal effects in a modular way. The parameters of these effects will be incrementally adjusted, and therefore, reflect the real values of low-level descriptions of the block. This will permit a fast but realistic bottom-up simulation of the whole system. The current methodology is based on the high level of reusability in the AMS IPs, achieved through modularity and parameterization.

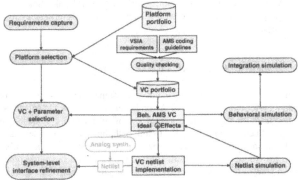

Figure 1. Mixed-signal reuse design flow

2.2   AMS IP classification

For the objective of building a formalized characterization currently available AMS IPs have been analyzed. In general, the sources can be divided in two classes: IP catalogues and IP providers. A set of attributes has been extracted for every IP. Then, this information has been elaborated with support of the results from [10] and [13], before a hierarchical AMS IP classification has been built with a set of attributes for every class and subclass.

The developed AMS IP classification is partly shown in Figure 2. The taxonomy is strictly hierarchical organized, and therefore, it can be represented by a tree. Each level in the taxonomy has special attributes associated. Also every item in the same level has special attributes. Each IP in the taxonomy inherits attributes from his mother class. For retrieval, the function or the characteristic of an AMS IP has to be specified as a combination of browsing within the taxonomy and providing additional input via attribute specification tables [31].

Every item in the *classification* has a set of attributes associated. Figure 3 shows as an example the set of attributes of a band-pass filter. Some attributes are common to every AMS IPs; for example, area, hardness, power consumption. But different classes of IPs require different attributes. For example, the order of the transference function for filters has as type enumeration and its enumeration values are e.g. 2nd, 3rd, 4th. This is not shared with non-filter AMS IPs.

Figure 2. AMS IP classification

	Attribute	Value
AMS IP Common Attributes	Name	CI5020xa
	Provider	Chipidea Microelectronics, SA
	Availability	Now
	Implementation	ASIC
	Input mode	Not available
	Output mode	Not available
	Technology	0.8 μm CMOS
	Retargetable towards	A
	Hardness	Hard
	Power down mode	No
	Power / current consumption	Not available
	Supply voltage	5 V
	Area	0.45 mm²
Filter Attributes	Architecture	Gm-C filter
	Load	Not available
	Order	4th order
	Offset auto-calibration	No
Band-Pass Attributes	Bandwidth	715 kHz
	Center frequency	21.4 MHz (range 15 MHz to 30 MHz)

Figure 3. Band-pass filter set of attributes

3. SYSTEM-LEVEL SPECIFICATION

The importance of documentation and specification during the first stages in the IP design process has been recently confirmed [17]. This well-recognized bibliographical source, although not containing explicit AMS considerations, has already shown the determinant role of specification for reuse and synthesis, and most of the rules provided are also valid for AMS macros in SoCs. Nevertheless, an extension of the methodology is needed regarding AMS particularities.

The success of using languages for digital hardware design during the last years, and the near standardization of languages for AMS descriptions [15, 28] suggest a central role for languages in the whole system conception. The capability of a language to state relevant information in order to guarantee a further use of a present implementation is clear, together with the hints that it can provide for simulation and synthesis.

Besides the characterization and classification described above, a current problem in SoC design is the different "wording" of each of the specialists, i.e. system architect, software engineer and hardware engineer. We argue that an AMS specialist is also necessary within the design team, at least until an actual seamless reuse and synthesis mechanism for SoC implementation will be running completely, which is the aim of our current work.

Thus, as a real working example, while the system architect uses C++, the hardware engineer might be working with VHDL, and the software engineer prefers writing embedded code in C. Therefore, an evident communication problem appears here. The solution to this problem represents one of the reasons for choosing SystemC [27], since version 2 of this language covers the area of the system architect and the ones of the software and digital hardware engineers. Other reasons for using SystemC are its system-level specification possibilities, enhanced with mechanisms such as the novel transaction level modeling [14], and the wide support from industry and academia.

Together with this scarcity in the coverage of models of computation [18] by standardized hardware-related languages, a lack of expressiveness for specification is found. Specification is not the same as modeling or purely functionality description. Specification means the particular conception of a system using modeling techniques. In the AMS area, standard languages for hardware design are not affording key specification aspects such as performance description or partition information.

3.1   Language notation

SystemC has demonstrated to be able to customize a wide range of models of computation [13]. In spite of this capability, an additional effort has to be made in order to formalize and directly support most of the models of computation. This reason has motivated the introduction of new developments regarding the improvement of AMS features in SystemC [9]. Therefore, early seizing in the extension of AMS capabilities to SystemC is performed in order to incorporate reuse and synthesis considerations regarding the specification of AMS systems within this language. The approach proposed here, will be viewed as an extension to the current design flow in SystemC. Thus, a set of elements is defined that aim at the extension of the specification expressiveness of the language, taking into account that the current lack of expressiveness is one of the key points which grows the reticence of analog designers to use languages for electronic circuit design.

The test bench for the evaluation and methodology inferences is a delta-sigma analog-to-digital data converter [25]. A complete delta-sigma converter represents a suitable system for our purposes, since it is a paradigmatic mixed-signal system example where simulation becomes a difficult matter [8]. The Figure 4 shows a first order delta-sigma modulator, which is the basic macroblock for a higher order converter. From this case study, together with some previous ideas taken from recent proposals like the aBlox notation reported in [7], some of the necessary characteristics for AMS specification have been extracted.

In SystemC, SC_MODULE is the language construct for the fundamental building block. A SC_MODULE specification must include the specification information, which is proposed to be contained in the following constructs:
- SC_DOMAIN indicates whether the SC_MODULE is analog, digital or non_determined
- SC_PERFORMANCE is an important construct for specification. It provides data that gives information for a design/architecture space exploration tool and/or for optimization.
- SC_REUSE is used for special information related to the reusability of the module, such as input and output impedance, technology or whether the ports are voltage or current values.

Inside the SC_CTOR construct, SC_CONTINUOUS would be introduced as another kind of process. It specifies that the module functionality will pertain to the analog domain. The future AMS SystemC language extensions need to support time domain simulation as well as frequency domain simulation. In the body of the module, support of functional specification at the Signal Flow Graph (SFG) level [26] is proposed, since SFG blocks suggest the structure of a system, and therefore effective CAD algorithms can be performed [6].

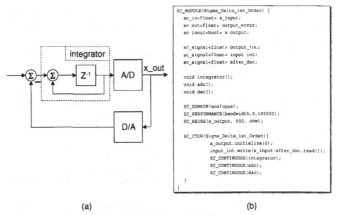

```
SC_MODULE(Sigma_Delta_1st_Order) {
    sc_in<float> x_input;
    sc_out<float> output_error;
    sc_inout<bool> x_output;

    sc_signal<float> output_int;
    sc_signal<float> input_int;
    sc_signal<float> after_dac;

    void integrator();
    void adc();
    void dac();

    SC_DOMAIN(analogue);
    SC_PERFORMANCE(bandwidth,0,100000);
    SC_REUSE(x_output, 500, ohm);

    SC_CTOR(Sigma_Delta_1st_Order){
        x_output.initialize(0);
        input_int.write(x_input-after_dac.read());
        SC_CONTINUOUS(integrator);
        SC_CONTINUOUS(adc);
        SC_CONTINUOUS(dac);
    }
}
```

(a)   (b)

Figure 4. Delta-sigma modulator (a) block diagram; (b) Code

3.2   Complete platform and further work

In order to accelerate the design cycles, analog designers should benefit
from the development of the languages, tools and methodologies that are
appearing. Nonetheless, more efforts have to be carried out to make them
suitable to the current design expertise. A direct way from the advances
reported to the handling of analog building blocks would be of great
importance for the analog design community. In any case, system-level
specification, as well as platform-based design and design reuse, has this
scope, which will be more largely improved with the definition of
macroblock constructs tackling functionality, performance and quality
aspects.

4.   CONCLUSION

As a consequence, the growth of integration capabilities and the new
market tendencies strengthen the necessity that System-on-Chip design
integrates an increasing amount of analog components. The complexity of
such systems for top-down design and full bottom-up verification makes it
necessary to face higher level of abstractions in the descriptions, and to
apply design reuse. The aspect of standardization is crucial for an effective

AMS reuse. This paper has summarized the current standards and guidelines in this area, and it has analyzed the different levels of abstraction of the digital, the analog and the mixed-signal domain.

The approach developed for AMS IP characterization has been presented, formalizing the current market situation, and furthermore, a design methodology based on SystemC, together with extended language constructs and a core example has been shown. This characterization is crucial for a proper IP management and usage, both at the intra- and inter-company level. The characterization is being extended to formalize descriptions at different levels of abstraction. The electronic design automation industry is paying a lot of attention to the field of analog synthesis. As a natural consequence, system-level descriptions will gain acceptance among the IP providers and integrators. Using a common language for system-level specification facilitates the communication among system, hardware and software engineers.

REFERENCES

[1] Barcelona Design, Prado:http://www.barcelonadesign.com, 2002.

[2] Cadence, Neocell: http://www.cadence.com/products/neocell.html, 2002.

[3] R. Castro-López et al.: "A Methodology for Building Mixed-Signal IP Blocks", Workshop on Mixed-Signal IP Blocks, Paris, 2002.

[4] H. Chang, L. Cooke, M. Hunt, G. Martin, A. McNelly and L. Todd, Surviving the SOC Revolution, Kluwer Academic Publishers, ISBN0-7923-8679-5, 1999.

[5] ComCad GmbH, Size: http://www.comcad-ads.de, 2002.

[6] Doboli, A. Nunez-Aldana, N. Dhanwada, S. Ganesan and R. Vemuri, "Behavioral Synthesis of Analog Systems using Two-Layered Design Space Exploration", in Proc. of the 36th DAC, 1999.

[7] Doboli and R. Vemuri, "A Functional Specification Notation for Co-Design of Mixed Analog-Digital Systems", in Proc. Design Automation and Test in Europe (DATE), Paris, France, 2002.

[8] Y. Dong and A. Opal, "An Overview on Computer-Aided Analysis Techniques for Sigma-Delta Modulators", in Proc. IEEE International Symposium on Circuits and Systems, 2001.

[9] K. Einwich et al, "SystemC extensions for mixed-signal design", in Proc. Forum on Design Languages (FDL 01), September 2001.

[10] E.N. Farag and M. I. Elmasry, Mixed signal VSLI wireless design, Kluwer Academic Publishers, ISBN 0-7923-8687-6, 1999.

[11] G. Forster, et al., "ACSYN – A Tool for Design and Characterization of Analog Circuits with Focus on RF Applications", in Proc. DATE 2001 User's Forum, Munich, pp. 148-152, 2001.

[12] J. Ginés, E. Peralías, N. Martínez Madrid, R. Seepold and A. Rueda, "A Mixed-Signal Design Reuse Methodology Based on Parametric Behavioural Models with Non-Ideal Effects", in Proc DATE, Paris, France, March 2002.

[13] T. Grötker, S. Liao, G. Martin and S. Swan, System Design with SystemC, Kluwer Academic Publishers, ISBN 1-4020-7072-1, 2002.

[14] S. A. Huss, Model Engineering in Mixed-signal Circuit Design, Kluwer Academic Publishers, ISBN 0-7923-7598-X, 2001.

[15] IEEE, Standard VHDL Analog and Mixed-Signal Extensions IEEE 1076.1, 1999.

[16] International Technology Roadmap for Semiconductors, http://public.itrs.net/, 2002.

[17] M. Keating and P. Bricaud, Reuse Methodology Manual, Second Edition, Kluwer Academic Publishers, ISBN 0-7923-8558-6, 1999.

[18] E. A. Lee and A. Sangiovanni-Vincentelli, A Denotational Framework for Comparing Models of Computation, Memorandum UCB/ERL M97/11, EECS, University of California, Berkeley, Jan. 1997.

[19] K. Liebermann et al., "Cambio: Compaction with Flexible Design Flow Integration", in Proc. DATE2001 Designers' Forum, Munich, pp. 62-64, 2001.

[20] N. Martínez Madrid and R. Seepold, "Virtual Component Reuse and Qualification for Digital and Analogue Design", in System-on-Chip Methodologies & Design Languages, P.J. Ashenden, J.P. Mermet, R. Seepold (eds.), Kluwer Academic Publishers, ISBN: 0-7923-7393-6, 2001.

[21] MEDEA+, Micro-Electronics Development for European Applications, www.medeaplus.org, 2002.

[22] V. Meyer zu Bexten et al., "ALSYN: Flexible Rule-Based Layout Synthesis for Analog IC's"; in IEEE Journal of Solid-State Circuits, Vol.28(3), pp.261-268, 1993.

[23] Neolinear, Neocircuit: http://www.neolinear.com, 2002.

[24] N. Faulhaber and R. Seepold, "A Flexible Classification Model for Reuse of Virtual Components", Chap. 3 in Reuse Techniques for VLSI Design, Kluwer Academic Publishers, ISBN 3-8265-3417-4, pp. 21-36, 1999.

[25] S. Norsworthy, R. Schreier and G. C. Temes, Delta-Sigma Data Converters, IEEE Press, New York, NY, ISBN 0-7803-1045-4, 1996.

[26] K. Ogata, Modern Control Engineering, Prentice-Hall, 1990.

[27] Open SystemC Initiative. SystemC 2.0 β-1 User's Guide, 2001: www.systemc.org, 2002.

[28] Open Verilog International, Verilog AMS Language Reference Manual 2.0, 2000.

[29] G. van der Plas, et al., "AMGIE : a synthesis environment for CMOS analog integrated circuits", in IEEE Transactions on CAD, vol. 20(9), pp. 1037-1058, 2001

[30] P. Schwarz, "Wiederverwendung von analog-digitalen Schaltungen", it+ti – Informationstechnik und Technische Informatik 44 (2002) 2. (in German), 2002.

[31] R. Seepold, "Reuse of IP and Virtual Components", in Proc. Design Automation and Test in Europe (DATE), Munich, Germany, March 1999.

[32] R. Seepold and A. Kunzmann (Eds.), Reuse Techniques for VLSI Design, Kluwer Academic Publishers, ISBN 0-7923-8476-8, 1999.

[33] R. Seepold and N. Martínez Madrid (eds.), Virtual Components Design and Reuse, Kluwer Academic Publishers, ISBN0-7923-7261-1, 2001.

[34] R. Seepold et al., "IP Design and Integration for System-on-Chip", Tutorial at Design Automation Conference (DAC), June 2002.

[35] TOOLIP Project (MEDEA+ A-511), http://toolip.fzi.de, 2002.

2

PART II: UML SYSTEM SPECIFICATION AND DESIGN

Chapter 8

A UML PROFILE FOR REAL-TIME SYSTEM MODELLING WITH RATE MONOTONIC ANALYSIS

Ian Oliver

Nokia Research Center, Itämerenkatu 11–13, Helsinki, Suomi/Finland.
E-mail: ian.oliver@nokia.com

Abstract   The uses of both the Unified Modelling Language and Rate Monotonic Analysis are increasing in the development of real-time systems. While both are addressing their usage it remains to be seen a clear, defined link between both UML and RMA. We present here a profile based upon the OMG's UML Real-Time Profile (Schedulability Analysis Subsection) that allows the placement of timing information and the subsequent generation of RMA models from UML models.

Keywords: UML, RMA, Real-Time, Schedulability, Timing, Validation

1.   INTRODUCTION

The complexity of embedded, real-time systems is increasing to a point where no longer 'traditional' techniques such as, for example, hand-optimising and writing of code or simple cyclic executive scheduling solutions are acceptable during a long-term development (Douglas, 1999; Selic et al., 1994; Awad et al., 1996).

For example a DSP processor in a modern mobile phone may play a number of roles such as audio en/de-coding, multimedia processing, GSM/GPRS functions etc. These demands coupled with the need for low CPU usage (to conserve battery power) plus load-balancing between concurrently running processes and in addition to this, other CPU devices means that we must have some way of assessing the potential the load upon the system early during design.

E. Villar and J. Mermet (eds.), System Specification and Design Languages, 87–106.
© 2003 *Kluwer Academic Publishers.*

It is in these kinds of environments that it is necessary to consider – at a very early stage in the development – the performance of the system (Binder, 2000) from the point of view of the model. Doing so means that we can guarantee and enforce a schedulable, low-power architecture throughout the software engineering process and avoid costly re-design to enable these properties during a later stage (Westland, 2002). In (Gomaa, 2000) how this analysis is performed is clearly demonstrated through rate monotonic analysis (Briand and Roy, 1999; Klein et al., 1993).

At Nokia we have developed OO based techniques for the development DSP software. We have also introduced the use of Rate Monotonic Analysis (RMA) (Klein et al., 1993; Briand and Roy, 1999) into the process to facilitate the assessment of CPU load during the requirements, analysis and design phases.

There now exists a UML profile (UML-RT) for real-time systems (OMG, 2002a) based upon the OMG UML semantics (OMG, 1999; OMG, 2002b). However this profile did not exist at the beginning of our project, thus it was necessary to construct a real-time profile ourselves at this time. During the course of the project the details of the UML-RT were released and were closely followed to ensure that any profile that we create would be compatible and based upon the OMG UML-RT profile (viz. Figure 1).

Although UML-RT does consider schedulability analysis by providing suitable stereotypes and tagged-values it does not define a mapping to RMA nor does it specify any relationships between the values and diagrams to support some method (Selic et al., 1994; Awad et al., 1996).[1]

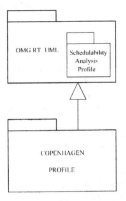

Figure 1. OMG UML-RT and Copenhagen RT profiles.

In this paper we discuss the syntax and semantics of the **Copenhagen Profile**, that we use in a DSP development environment to annotate models with timing information. We show how this profile is then used to map the UML models to RMA models. We briefly discuss the available tool support and outline the future work in this area.

2.   COPENHAGEN PROFILE

The use of the UML for the specification of real-time systems is desirable for similar reasons to that of 'normal' systems; UML provides a graphical way of understanding the structure of the system and provides a common language between developers. However real-time systems have characteristics that requires the use of a profile of the UML tailored towards the needs of real-time modelling. We have developed our Copenhagen Profile such that it is now a sub-profile of the UML-RT as shown in Figure 1.

We use the convention for our additional extensions such that they are prefixed with the designation **COP** similarly to the UML-RT stereotypes and tagged values from the schedulability analysis subsection which are prefixed with **SA**.

2.1   Time

We use the UML-RT model of time in our profile. We do however state that all timing values must be specified in some unit of time which by default is either milliseconds or CPU cycles. A function describing the relationship between CPU cycles and actual physical time must be specified when using this profile.

The execution timing figures are all relative to the "previous message send". The use of absolute timing figures which are relative to the "initial message receipt" proved not be useful in the context we were using the timing values and are not considered in the profile as it stands here.

2.2   Modelling overview

We introduce a weak process/methodology which states that one initially uses class diagrams to outline the structure of the system and the timing values. These timing values are then progressively refined through the use of more detailed models with the sequence and state diagrams, viz:

Class diagram $\sqsubseteq$ Sequence diagram $\sqsubseteq$ State diagram $\sqsubseteq$ Method

The UML class diagram is used initially to specify the structure and the basic or *best guess* timing figures for the system.

The use of *best guess* timing figures means that it is possible to enforce upper-bounds on timing values at the system structure level. Refinements to those component must then conform to the already specified upper-bound figures (Morgan, 1990). A useful 'heuristic' to employ is to allow an +/–10% variation on these best-guess figures.

When development is made by a number of subgroups each working on particular components within the system this has proved to be a useful idea, especially when emphasis is made on reducing the figures as much as possible in the refinement.

Particular events in the system are specified using the sequence diagram which also serves (in conjunction with the collaboration diagram) as a way of specifying particular instances of the model. Timing values present in sequence diagrams are refinements of those that appear at the class diagram level.

Sequence diagrams are refined by state diagrams. One area where attention must be paid is that we make a link between the messages on the sequence diagram to those in that trigger transitions in the state diagram.

Deployment diagrams are used to specify a logical model of the target hardware. This model is usually no more complex than denoting which are processing elements and which are device elements. All elements in the deployment diagram have a rate parameter associated with them which acts as a coefficient to all timing values associated with that device. This enables the modeller to 'change' a processor for a faster or slower device.

Collaboration diagrams are used to specify particular scenarios or modes of operation. They are also used to track the interaction between objects and also document the threading model.

2.3   Task based modelling

We concentrate primarily on event driven systems (Pneuli and Manna, 1993; Selic et al., 1994; Nissanke, 1997) where the focus of control is made around active objects which have their own thread of control. When specifying the structure of the system we use the stereotypes – *COPactive* and *COP_PDC* (PDC or passive data class) to denote this. The semantics are slightly more detailed when one takes into consideration the underlying RTOS but basically an active class will have its own message handler and thread within the operating system environment.[2] Consider the model in Figure 2.

Here we have a trivial, single task system where the best-guess timing value for the active class C is 200ms and the triggering message m for some operation upon an instance of C has a period of 1000ms and deadline of 750ms.

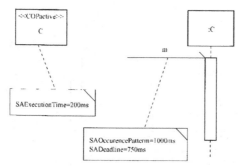

Figure 2. Trivial single task UML model.

It is assumed that normal RMA consistency rules always apply such that the deadline must be less than the period etc.

This would then map into an RMA model consisting of one task with said period, deadline and execution times of 1000ms, 750ms and 200ms respectively.

The situation is similar for that of two tasks, for example in Figure 3 the RMA model would consist of two RMA tasks with said period, deadline and execution times. The order of message processing would depend upon the priority of the messages and the underlying scheduling mechanism although *Run-to-completion* semantics are always assumed.

We now consider the case where two tasks may interact, for example in Figure 4 from which we obtain the RMA model shown in Figure 5.

Here we also refine class C's best-guess timing value with timing values on the sequence diagram. The sequence diagram now reads:

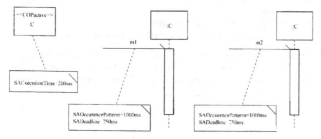

Figure 3. Trivial two task UML model.

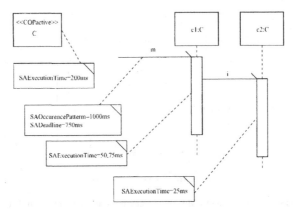

Figure 4. UML model of two interacting tasks.

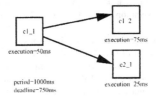

Figure 5. RMA model of two interacting tasks.

1   Message m is received by object <u>c1:C</u>
2   <u>c1:C</u> performs 50ms of execution
3   <u>c1:C</u> sends message i to object <u>c2:C</u>
4   IN PARALLEL
    (a)   <u>c1:C</u> performs 75ms of execution
    (b)   <u>c2:C</u> receives message i and performs 25ms of execution

The order in which execution takes place depends on the underlying execution model and the priorities of the messages and threads if specified. In a single processor environment the task structure would then map as follows:

Model	Execution
$c1_1 \rightarrow (c1_2 \parallel c2_2)$	$c1_1 \rightarrow c1_2 \rightarrow c2_2$
	$c1_1 \rightarrow c2_2 \rightarrow c1_2$

2.4   Sequence and state diagrams

In order to refine the model down to state diagram level it is necessary for us to define a relationship between the sequence and state diagrams. This is achieved by stating that messages (to which a task reacts) are the same messages to which a state diagram transition is triggered.

In Figure 6 we show a situation where refinement has taken place. We have a message m that triggers some execution in object[3] :C which takes 50ms. The refinement for message m is made in class C's state diagram which states that when m is received while in state S1 the processing takes 25ms and when while in S2 processing takes 45ms.

In Figure 7 we show the same model as in Figure 6 with the addition of state information to the sequence diagram. Using this information we can now obtain the correct timing value for a given message.

When the state is not directly specified on the sequence diagram then the default behaviour is to choose the highest, refined timing value, which in this case would have been 45ms from the S1-to-self transition.

On states also may be specified any entry and exit times associated with those states. This is achieved by using the *COPstateentry* and *COPstateexit* tags. Both tags take a timing value as the argument. This argument may be for 0 units of time. The relationship between the execution times and the state entry/exit times can be seen in Figure 8.

2.5   Method calls

The rôle of passive data classes (PDC) is to encapsulate some kind of functionality or data. In the DSP development PDCs are used for holding

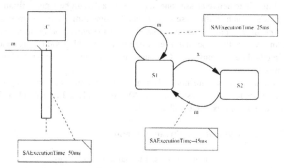

Figure 6. Sequence and state diagrams (ambiguous).

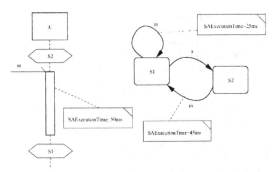

Figure 7. Sequence and state diagrams (unambiguous).

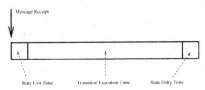

Figure 8. State entry/exit times and transition execution time.

certain kinds of function, for example; fast fourier transforms. This kind of encapsulation is useful and necessary in the cases where it is desirable to move blocks of data or functionality to certain kinds of device, eg: slow vs fast memory devices.

Similarly to active classes one may specify a default best-guess timing value on PDCs, which in this case would specify something such as: "*Any function written in this class must conform to this maximum timing upper bound*". Normally it is the case that after refinement the timing values are placed on the method calls themselves. Of course method calls may be written on active classes as well.

Calls to methods are normally made by a function call meaning that the caller waits for the processing in the PDC to complete. For example in Figure 9 we can see an example call to a PDC.

2.6   Deployment and thread allocation

The deployment diagram is used to build a logical model of the hardware on which the model runs. A deployment model is necessary in order to specify

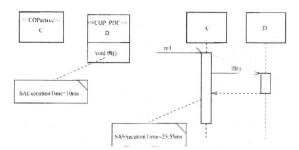

Figure 9. Method call to a passive data class.

the processor rate, device acquisition and deacquisition times and the function
that converts processor instruction cycle times to actual timing values.

We define two stereotypes for use in the deployment diagram

- COP_CPU
- COPdevice

The CPU stereotype admits the following tagged values:

- COPcontextswithctime
- COPrate
- COPcpucycletime

The device stereotype admits the following tagged values

- COPaqtime
- COPdeaqtime
- COPrate

An example deployment diagram can be seen in Figure 10.

To relate a class to a device it is necessary to introduce two tagged values:
COPonDevice and COPonCPU, to passive data classes and active classes
respectively. These tagged values take as their values any suitable devices
or CPUs respectively present in the deployment diagram.

The *onCPU* tag for active classes allows direct association with a particular CPU element (Kabous and Nebel, 1999). It is often necessary to allow
a threading structure to be built upon a CPU. To achieve this we introduce
the tags COPonThread and COPhasThreads to active classes and CPUs respectively. We also introduce tag COPthreadPriority to CPUs which takes a string
of parameters in the form [threadname,threadpriority]*. The meta-model for
this structure can be seen in Figure 11.

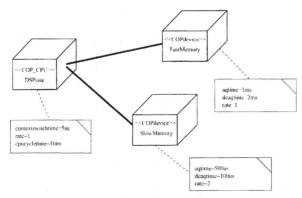

Figure 10. Example deployment diagram.

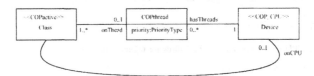

Figure 11. Threading model.

If an active class has both the COPonCPU and COPonThread tags then the value in the COPonCPU tag takes priority and that active class runs on the given CPU with *its own thread of control.*

2.7   Modes

A mode represents a particular state of operation of a system, for example a DSP in a mobile telephone could be in one of the following modes: boot, synchronisation, idle or traffic – relating to boot-time, acquiring the signal from the aerial, waiting for a call or some other processing and actually performing the A–D processing needed for communication respectively.

The configuration of objects, relationships and events varies with each mode. In our DSP analysis the traffic mode has a number of sub-modes of operation representing different event arrival patterns.

To describe the object configuration of a mode we may use either the collaboration or object diagrams – in reality we choose the collaboration diagram

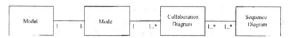

Figure 12. Mode modelling structure.

as it is possible to show the event interactions between objects, although the only significant elements are the objects themselves.

For each collaboration diagram we then associate a number of sequence diagrams which describe the events that occur during that mode. A sequence diagram may be shared amongst a number of collaboration diagrams. The structure of this can be seen in Figure 12.

We do not consider how the system behaves during mode changes (Pedro and Burns, 1998). In our experience and systems the mode changes are rare compared to the normal operation of the system and the change does not affect the overall timing of the system.

3.   MAPPING UML TO RMA

In this section we present the mapping from the UML model to the RMA model.

While no specific notation for noting RMA models exists, we havechosen to default on the notation used in the TriPac RapidRMA tool which uses a box to denote a task and arrow to denote dependency. The arrow is written in the direction of flow of control. The RMA model can be considered as a directed, non-cyclic graph. We supplement the RMA notation by writing the letters a, s, r or f (asynchronous, synchronous, synchronous return and function call respectively) alongside the arrow to denote the type of message being sent between tasks.

3.1   Class diagrams

Class diagrams present the trivial case when mapping to the RMA model. The class and its best-guess timing value simply map to a single RMA task with execution time as that of the best-guess value.

3.2   Mapping sequence diagrams

We must now consider four basic cases for the mapping of the UML to the RMA model. These cases are based around the sequence diagram which is capable of showing the relationship between the execution and calls between objects.

The first case we consider is that of a simple execution block. In Figure 13 we can see an incomming message triggering a block of execution. Note the naming convention between the UML and RMA models.

In the second case in Figure 14 we consider a function call message between two objects. It is normally the case here that all components of the execution are bundled together to form a single RMA task.

In the third case in Figure 15 we consider the asynchronous message. It may be the case here that the two RMA tasks related by the function call message can be joined as in the case in Figure 14. However this is really decided by the underlying execution mechanism discussed later.

The fourth case in Figure 16 we show a synchronous message between two objects. The RMA formalism does not support the callback of synchronous messages to the sending tasks fully. However in our experience the following mapping captures the worst-case scenario successfully in most cases (Kopetz, 1998). Note that the execution block a2 can not be executed until execution block b has at least started to execute.

Figure 13. Simple execution block mapping

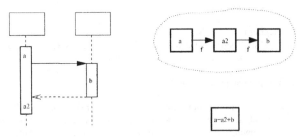

Figure 14. Function call message mapping.

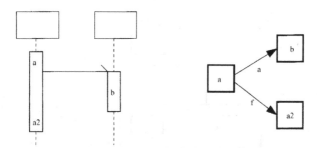

Figure 15. Asynchronous message mapping.

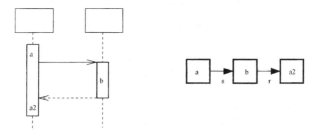

Figure 16. Synchronous message mapping.

3.3   State diagrams

The state diagram's timings are a refinement of the sequence diagram's timings. Here the transition between states contains similar information to that on the focus of control of a sequence diagram. This can be better shown in the following example. In Figure 17 we can see the class and state diagrams responsible for the structure and behaviour of the system we are modelling – primarily however we are interested in the state diagram's timings and especially the transition timing of 8ms,16ms. This timing figure must be a refinement of the sequence diagram timing figure for that message between those two particular states.

In Figure 18 we can see the sequence diagram detailing the execution involved upon receipt of message **t** between states **s1** and **s2**. Given this sequence diagram we obtain the RMA task graph as shown in Figure 19. If we now take into consideration the state diagram from Figure 17 we obtain

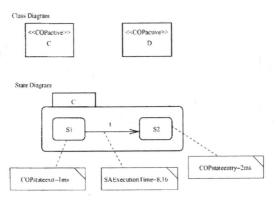

Figure 17. Simple class and state diagrams.

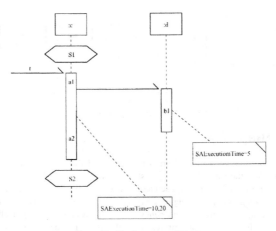

Figure 18. Simple sequence diagram.

the RMA task graphs shown in Figure 20. Note that we can treat the entry and exit times of the states and being analogous to function calls. Note also that the total timing of the state diagram transition plus state entry and exist times must be less than or equal to the total timing given on the sequence diagram.

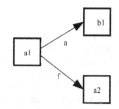

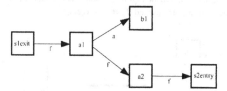

Figure 19. Task model after sequence Diagram (from Figure 18).

Figure 20. Task model after state diagram refinement.

Note in Figures 19 and 20 we may see the RMA task graph collapse with the unification of those tasks joined via function call messages as per Figure 14.

3.4   Devices and semaphores

We have discussed how the task specifications map to the RMA world but not explicitly how the hardware or deployment mapping is made. We devised no notation for hardware in RMA as the information contained in the model was normally incorporated into the RMA task timing information.

Taking the deployment diagram in Figure 10 into consideration we map to the following RMA logical devices:

- RMA_logical_cpu
- RMA_logical_device

In addition to this must create corresponding *logical semaphores* in order to preserve the fact that the logical cpus/devices can only support one executing task/access at any one time. In most RMA tools this is assumed for CPUs. It is possible to support non-binary-semaphores so that multiple access to devices is possible – this situation is however rare in our experience and normally we do not support this mode of operation for logical semaphores.

3.5   Threading model

The threading model presents an interesting situation regarding how
the RMA task graphs are implemented (Oliver, 2000). Normally all RMA
tasks can be unified when not connected via an asynchronous message – at
least this is true when considering some of the underlying message passing
systems in use in certain kinds of software we have used and also the run-
to-completion semantics used in the implementations generated by some CASE
tools.

Generally the cases listed in Figure 21 hold. The dotted lines correspond
to thread allocation and are named T_n respectively.

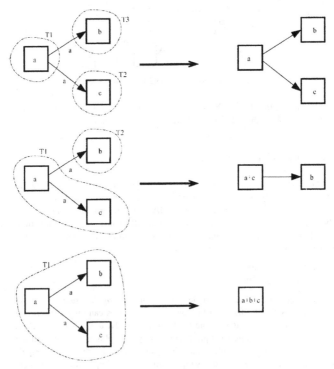

Figure 21. RMA task graph implementations.

4.   TOOL SUPPORT

For this kind of analysis to be successful it is imperative that some kind of tool support is available that supports this profile and the mapping it describes between the UML and the RMA worlds.

As part of this work support was provided via a plug-in module for Rational Rose Real-Time[4] developed by TriPac which integrated with the already existing TriPac RapidRMA tool.

The tool integration was such that the workings of the RMA toolkit were effectively hidden from the UML CASE tool user such that all integration and feedback was via UML diagrams, plus some graphs showing execution profiles – for example see Figure 22.

A small, example case-study showing the use and capabilities of this tool integration can be found in (Oliver, 2001).

5.   SUMMARY

In this paper we have described our chosen annotations and semantics for a sub-profile of the UML-RT profile which we have tailored specifically for rate monotonic analysis features. We have attempted only to add to the features discussed in the UML-RT profile (and in particular the schedulability sub-section), thus making our profile directly compatible with the proposed OMG UML-RT profile.

The profile is based upon a task based modelling methodology which we apply for the development of – primarily – embedded DSP systems. Currently used is an internally developed method based upon ROOM and Octopus which encourages the developer to take into consideration timing considerations. Indeed for the PDC development obtaining these values is relatively easy.[5] However timing analysis of some of the control oriented functionality has been more problematical. At least we strive to present the developers with upper bounds on the timing characteristics of their code.[6]

The ability to be able to predict the timing characteristics of the system at modelling stage and identify problem areas or areas where slack does exist has proved extremely valuable in assessing the architecture of the DSP software and planning how to integrate extensions and optimise code.

It must be said that these benefits do not come automatically but only after experiment, trial and error of these analysis techniques. The biggest problems have been initially obtaining the timing values and then understanding the results. As with any technique once some benefits have been shown, further introduction and use of the technique becomes more widespread and easier.

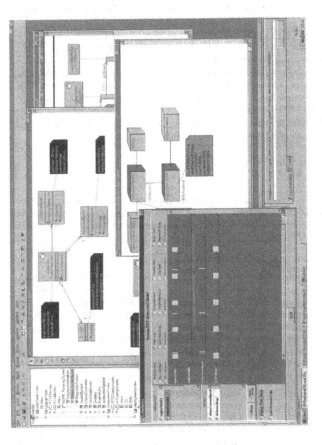

Figure 22. Rational Rose and RapidRMA integration snapshot.

6. ACKNOWLEDGMENTS

Martin Astradsson and Paul Groskopf at Nokia Mobile Phones and Ben Watson at TriPac, Dallas for valuable contributions to this work.

7. COPENHAGEN PROFILE SUMMARY

Table 1. Copenhagen profile stereotypes

Stereotype	Element	Notes
COPactive	Class	Class with own thread of control
COP_PDC	Class	Class encapsulating data or functionality
COP_CPU	Device	Denotes a processing device, e.g.: CPU
COPdevice	Device	Denotes a non-processing device, e.g.: Memory
COPthread	Class	Denotes a thread or similar mechanism

Table 2. Copenhagen Profile Tagged Values

Tag	Element	Notes
COPcontextswitchtime	COP_CPU	Processor context switch time
COPcpucycletime	COP_CPU	Time of 1 cpu cycle
COPrate	COP_CPU, COP_PDC	Rate of device
COPaqtime	COP_PDC	Aquisition time
COPdeaqtime	COP_PDC	Deaquisition time
COPonDevice	COP_PDC	Denotes on which device the PDC is implemented
COPonCPU	COPactive	Denotes on which CPU the Class is implemented
COPonThread	COPactive	Denotes on which thread the Class is implemented
COPhasThreads	COP_CPU	Denotes which threads sun on the CPU
COPstateentry	States	State entry time
COPstateexit	States	State exit time
SAExecutionTime	*any period of execution*	Execution Time in cycles or time
SAOccurencePattern	Messages	Rate at which message occurs
SADeadline	Messages	Deadline for complete processing of message

NOTES

1. It is also our belief that the UML (and UML-RT) must be tailored or configured to individual projects' needs through the UML's extensibility mechanisms.
2. OSE, Rational Rose Real-Time etc.

3. Anonymous object.
4. A plug-in for I-Logix Rhapsody has been prototyped as well.
5. Assembly language statement counting usually.
6. One could say that it becomes a matter of skill and pride improve efficiency.

REFERENCES

Awad, Maher, Kuusela, Juha, and Ziegler, Jürgen (1996). *Object-Oriented Technology for Real-Time Systems. A Practical Approach Using OMT and Fusion.* Prentice-Hall.

Binder, Robert V (2000). *Testing Object-Oriented Systems – Models, Patterns and Tools.* Addison-Wesley. 0-201-80938-9.

Briand, Louc P. and Roy, Daniel M. (1999). *Meeting Deadlines in Hard Real-Time Systems – The Rate Monotonic Approach.* IEEE The Computer Society.

Douglas, Bruce Powel (1999). *Doing Hard Time. Developing Real-Time Systems with UML, Objects, Frameworks and Patterns.* Addison-Wesley. 0-201-49837-5.

Gomaa, Hassan (2000). Designing Concurrent, Distributed and Real-Time Applications with UML. Addison-Wesley. 0-201-65793-7.

Kabous, Laila and Nebel, Wolfgang (1999). Modelling hard real time systems with uml, the ooharts approach. In France, Robert and Rumpe, Berhard, editors, *UML'99 – The Unified Modelling Language,* Lecture Notes in Computer Science 1723, Springer, pp. 339–355.

Klein, Mark H., Ralya, Thomas, Pollak, Bill, Obenza, Ray, and Harbour, Michael Gonzalez (1993). *A Practitioner's Handbook for Real-Time Analysis: Guide to Rate Monotonic Analysis for Real-Time Systems.* Kluwer Academic Publishers.

Kopetz, Hermann (1998). *Real-Time Systems: Design Principles for Distributed Embedded Applications.* Kluwer Academic Publishers.

Morgan, Carroll (1990). *Programming from Specifications.* Prentice-Hall.

Nissanke, Nimal (1997). *Realtime Systems.* Prentice Hall.

Oliver, Ian (2000). Integrating schedulability analysis with uml models of real-time systems. In Smolyaninov, Alexander and Shestialtynov, Alexei, editors, *WOON2000, 4th International Conference on Object-Oriented Technology, St. Petersburg, Russia,* pp. 70–81.

Oliver, Ian (2001). An example of validation of embedded system models described in uml using rate monotonic analysis. *Proceedings of SIVOES2001 at ECOOP2001, Budapest, Hungary.*

OMG (1999). *Unified Modelling Language Specification (draft).* Object Management Group, version 1.3r9 edition.

OMG (2002a). *Response to the OMG RFP for Schedulability, Performance and Time.* Object Management Group, revised submission edition.

OMG (2002b). *Unified Modelling Language Specification (Action Semantics).* Object Management Group, version 1.4 (final adopted specification) edition. OMG Document Number ad/02-01-09.

Pedro, P. and Burns, A. (1998). Schedulability analysis for mode changes in flexible real-time systmes. In *Proceedings, 10th Euromicro Workshop on Real-Time Systems, Berlin, Germany.* IEEE Computer Society, pp 17–19.

Pneuli, Amir and Manna, Zohar (1993). Models for reactivity. *Acta Informatica* 30(7): 609–678.

Selic, Bran, Gullekson, Garth, and Ward, Paul T. (1994). *Real-Time Object-Oriented Modelling.* Wiley.

Westland, J. Christopher (2002). The cost of errors in software development: evidence from industry. *The Journal of Systems and Software* 62(1): 1–10.

Chapter 9

SUPPORT FOR EMBEDDED SYSTEMS IN UML 2.0

Morgan Björkander[1], Cris Kobryn[1]
[1]Telelogic

1. INTRODUCTION

The Unified Modeling Language (UML) [8] is currently undergoing a major revision process that will eventually lead to the creation of UML 2.0. This process was initiated in 2000 in order to address the various shortcomings of the original language, and to keep it current with new technology trends such as component-based development. The new revision is currently due by the end of this year. While there are several active proposals for UML 2.0, they are largely complementary, and collaborations are underway to merge the submissions. However, for the purpose of this paper we concentrate on the submission from the largest submission team [10, 11], and note that since the revision is not yet finalized the examples given here are tentative.

In this paper we examine structural and behavioral constructs that have traditionally been used within the embedded systems industry to develop real-time systems and are now being incorporated into UML for more general use. These are proven scalable concepts that already exist in languages such as SDL [1], MSC [2], and have also appeared in the UML-RT profile [9]. Additionally, executable modeling languages such as SDL have had a primary role in the creation of an Action Semantics for the UML [4], a related specification that we also explore in this paper.

Before we start looking at new features, it is worth noticing that the UML 2.0 designers are striving to ensure backward compatibility with UML 1.x,

E. Villar and J. Mermet (eds.), System Specification and Design Languages, 107–118.
© 2003 *Kluwer Academic Publishers.*

meaning that most of the existing language capabilities are preserved. From a user's perspective, this helps protect current investments in UML, and makes it easier to migrate to UML 2.0.

2.   ARCHITECTING SYSTEMS

One of the biggest challenges using UML 1.x is how to model large, complex systems using the available concepts. Although most UML 1.x modelers who are familiar with object-orientation find the *Class* construct intuitive and flexible, many have encountered difficulties in applying more advanced constructs, such as *Subsystem*, and *Component* to structure large systems. As an example, the creators of the UML profile for EJB™ [3] found that *Subsystem* was a better match for representing Enterprise JavaBeans™ components than *Component*. In short, the challenge has been to describe software architectures of different flavors, where different domains use terminology inconsistently and make subtle, but important distinctions in semantics.

2.1   Components and Interfaces

To address this in UML 2.0 we rely on some common techniques that are well tested involving compositional building blocks that promote the use of contracts between entities of a system. While the class concept remains largely untarnished, the definition of component is significantly updated. (The existence of subsystem is currently being argued, but at the time of this writing its ultimate fate had not yet been determined.) A component in UML 2.0 thus represents a building block that encapsulates its structure and behavior. Most component-based development techniques rely heavily on interface-based design, and from the outside a component is viewed as a "black box" whose services are exposed through the use of interfaces. When designing the component, it is necessary to look inside the component, where its actual implementation in terms of structure and behavior is described (the "white box" view of the component).

While UML 1.x only had a shorthand notation for describing *provided* interfaces in the lollipop symbol, this has in UML 2.0 been expanded to also allow a shorthand notation for *required* interfaces. This way, it is possible to develop each building block as a stand-alone entity whose dependencies to other building blocks can be represented entirely through interfaces. Components tend to be active in real-time systems, i.e., they have their own thread of control. In addition, since communication in real-time systems tend to be asynchronous rather than synchronous it is necessary to be able to represent signals (or rather reception of signals) in interfaces. For this, the same notation as for operations is used, but preceded with the keyword «signal».

2.2 Component ports

A component may provide a large set of services, not all of which are suitable for every user of the component. For this reason it is possible to describe interaction points, or *ports*, of a component that represent different views of how the component may be used. Each port may have a set of required and provided interfaces and can be separately addressed. A port sits at the boundary of a component, and gives users an access point to its implementation. Similarly, it provides the internal structure of a component with an access point to the environment of the component. It can therefore be thought of as an opening in the shell of a component where it is possible to look in from the outside and out from the inside. All that you see, however, are the interfaces supported by the respective port. The primary purpose of a port is to act as a connection point when components are connected to each other as parts of other components, which we look at next.

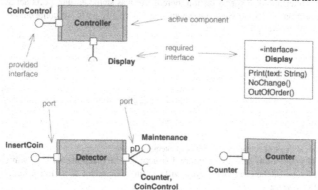

Figure 1. Definitions of a set of components with ports and interfaces

2.3 The internal structure of a component

The granularity of a component is not set; the same component can, depending on the context where it is used, represent an entire system or only a minor part of a larger system. The design of a component may therefore be arbitrarily complex, ranging from directly implementing the provided services in terms of behaviors such as state machines or activities to having an internal structure. In the latter case, a component is decomposed into smaller *parts*, each represented by another component or a class. There is a compositional aspect here, where each part is owned by the containing component; this implies that when an instance of a component is created, all its parts are also created. Likewise, when the instance is later terminated, all parts are also terminated.

The internal structure is made up of parts that are connected to each other using *connectors*. Each part represents instances that are going to be created of a particular class or component when an instance of the containing component is created, and a connector represents a communication path between two parts. A defined component may be used as a part in many different contexts, and in each of those contexts it may be connected differently. Connectors therefore differ from associations in that they are contextual; while an association is always valid in any context, a connector is only applicable in a specific context. It is only possible to connect parts of components that have matching interfaces, i.e., if one component requires an interface it can only communicate with other components that provides the same interface or a specialization thereof. As was stated earlier, ports act as connection points between components, and normally parts are connected to each other through their ports.

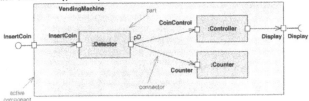

Figure 2. Usage of components as part of an internal structure

If you "zoom out" of an internal structure, which represents the "white box" view of the component, you will see the component as a "black box," and vice versa.

2.4   Behavioral ports

Even if a component has an internal structure it may still have behavior of its own that for example implements certain interfaces or controls the internal structure. Ports that connect directly to the behavior of a component rather than to its internal structure are called behavioral ports. The distinction is normally only made in the "white box" view of a component since users should not know about the component's internal design, and is shown notationally by attaching a state symbol to the port.

Behavioral ports are used primarily when parts of an internal structure need to communicate directly with the containing component, but also when the containing component should not delegate the implementation of a service to one of its parts. In the former case, the behavioral port normally has protected visibility, while in the latter case it needs public visibility. The behavior of the component may be represented through any behavior, including state machines, activities, and interactions.

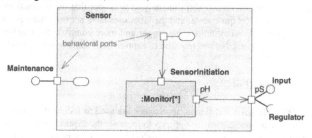

Figure 3. A component with a public and a protected behavioral port

2.5 Protocols and contracts

The interfaces of a component do not give the full story about how to use the component, since they do not tell you much about how the component works. This other part of the story can be accomplished in many ways. Perhaps the simplest is to go ahead and examine the implementation of the component, but this is often cumbersome or even too complex, and also breaks the supposed encapsulation of the component.

Instead, there are better ways to specify how a client may use a component, and how components may interact with each other. *Interactions* are ideal to specify the communication that may (or may not) occur between any number of components. Alternatively, or in addition, it is possible to attach a protocol state machine to an interface to indicate how the interface may be used, for example the order in which signals may be received or its operations be invoked. A protocol state machine attached to a port indicates the valid interactions that may occur at that port. Below, we elaborate on how interactions work.

3.   USING INTERACTIONS TO DESCRIBE PROTOCOLS

Interactions, or sequence diagrams, constitute some of the most powerful concepts of UML, and have many different purposes. We have already mentioned that they can be used for showing the valid interactions between components, and this is in fact their primary purpose today: to show traces between instances of classes or components. However, interactions are also eminently suitable for capturing and explaining requirements of a system or parts of the system, and for expressing tests and test suites that go hand in hand with the requirements. Unfortunately, it turns out that the interactions of UML 1.x does not quite scale, and the latter uses have not been realized to their full potential. As systems grow bigger and more complex, the number of sequence diagrams that are required to capture their functionality quickly become staggering.

3.1   Variations

In UML 2.0, several different approaches are used to reduce the number of required sequence diagrams, but also to increase the expressive power of interactions. The first of these is the introduction of variations in a sequence diagram, which are accomplished by what is called combined fragments. The notation for a combined fragment is a frame encapsulating a number of lifelines and the messages between them. The variation modeled by a particular combined fragment is determined through the use of operands, which are shown in the upper left corner of the frame. If there is more than one operand, these are separated by a horizontal dashed line.

In the example shown in Figure 4, an alternative is used, which indicates that after the *VendingMachine* has received the *Insert* message from a *User*, it may respond either with a *Display* or a *RejectCoin* message. In order to express this kind of functionality in UML 1.x it would have been necessary to rely on two sequence diagrams. While it is not shown here, it is possible to associate a guard with each alternative. Other variations include:

- loops (loop): the messages within the variation are repeated a number of times,
- optionality (opt): the messages within the variation are optional,
- filters (filter): not all messages are shown within the variation, i.e., some messages are filtered out,
- impossible occurrences (neg): sequences of messages that must not occur.

Other predefined variations are for example assertions (assert) of sequences that must occur, critical regions (region), strict sequencing across all lifelines (strict), and parallel sequences of messages (par). Traditionally, the primary strength of sequence diagrams has been their simplicity, which means that virtually anyone can understand them. Care must be taken with variations, because if they are overdone the sequence diagrams may easily become unintelligible.

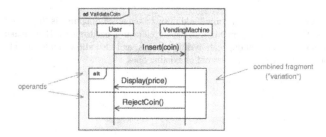

Figure 4. Showing variations in a sequence diagram

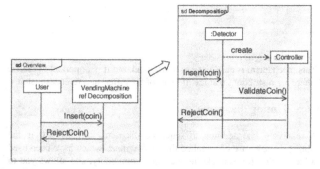

Figure 5. Decomposing a lifeline

3.2   Decomposition of lifelines

Just like it was possible to decompose components into parts, there is a corresponding capability that is defined for sequence diagrams, where a component lifeline can be decomposed into a new sequence diagram that shows the interactions between its parts. This way, it is possible to first show the interactions with a component as a whole, and the zoom into the details of how those messages are dealt with internally by the component's parts.

In Figure 5, the lifeline for the *VendingMachine* component is decomposed into another sequence diagram called *Decomposition*, where we can see the how the message *Insert* that was sent to the *VendingMachine* is actually being delegated to a *Detector* that is part of the *VendingMachine*.

3.3   References to other interactions

One of the most important additions to sequence diagrams is the ability to reference other interactions, so called interaction occurrences. These make it possible to avoid duplication of sequences, and also allow you to quickly put together new sequences out of existing ones.

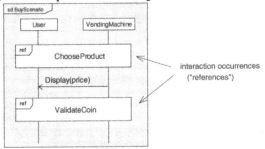

Figure 6. Referencing other interactions

This mechanism is quite easy to understand, but it is important to realize that there are restrictions that the lifelines of the different references match.

3.4   Timing diagrams

A relatively late addition to sequence diagrams in UML 2.0 is the possibility to make use of so called timing diagrams, which have been used for a long within for example the hardware community, and allows you to model the state changes over time in sequence diagram. Of particular importance is that this kind of diagram allows you to see how different objects are related to each other in terms of states.

Note that these diagrams are traditionally tilted 90 degrees to show time on the horizontal axis and state changes on the vertical axis, whereas normal sequence diagrams show time increasing on the vertical axis.

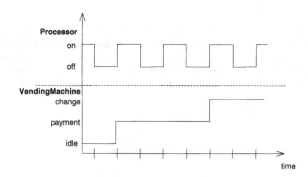

Figure 7. Timing diagrams

4. SPECIFYING SYSTEM BEHAVIOR USING STATE MACHINES

In real-time systems, state machines are almost as important as sequence diagrams when it comes to expressing the behavior of classes and components. State machines inherently deal with asynchronous communication, and are ideal to capture protocol specifications. In UML 2.0, state machines have been somewhat simplified and easier to use, and at the same time they have been made more scalable. There was furthermore some discussion about protocol state machines in UML 1.4, but the concept was never fully worked out. In UML 2.0 this has been remedied.

Some of the more important additions to state machines include the capabilities of having simple and composite states of a state machine. This provides a way to express state machines in a very scalable way, where essentially any state can be subdivided into any number of states, which in turn may be composite. In addition, the concepts of entry and exit points have been added that can be used to determine different execution routes through a composite state. This allows a single composite state to express behavior that would otherwise require multiple composite states and make the model much harder to understand and in some cases require duplication of information.

4.1   Action semantics

The addition of an Action Semantics for the UML [4] is largely inspired by the way SDL went from being a pure specification language to more and more often being used as a programming language; the action semantics precisely defines the semantics of executing actions such as assignments, loops, and decisions as part of for example transitions of a state machine.

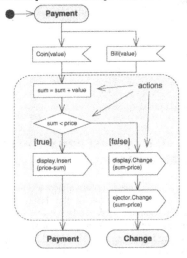

Figure 8. Actions as part of a state machine

In Figure 10, the highlighted area represents the area that is affected by the action semantics. Each of the symbols represents an action, in this case an assignment, a decision, and a few call actions. The action semantics apply anywhere actions are used, such as shown here in state machines, in activities, or even in interactions. This view of a state machine is commonly referred to as transition-centric, since it focuses on the actions that are part of for example a transition; it complements the more commonly used state-centric view when expressing detailed behavior.

The ability to describe system functionality at a higher abstraction level than in an ordinary programming language such as Java or C++ makes it possible to execute UML models, which further means that it becomes possible to verify systems at a much earlier stage in the development lifecycle, to automate testing against UML models, and to apply different forms of schedulability and performance analysis to the model. An additional benefit is that the model becomes not only platform independent, but also target language independent. Given appropriate transformation rules, it is possible to swiftly change an implementation from one programming language to another, and the generated code can also be easily optimized for different situations, for example to minimize footprint or to maximize speed of execution. Since the full behavior can be expressed directly in the model, it should never be necessary to modify the generated code.

The Action Semantics for UML is a complete update of the UML specification itself, but is based on UML 1.4. Rather than wait for a separate revision process to adapt it to UML 2.0, the action semantics is being integrated into the language as part of the new revision proposal. Neither the Action Semantics for the UML nor UML 2.0 itself specify a notation to be used for action statements; the example shown above uses a made up notation that is based on Java syntax.

5. RELATED STANDARDS AND NEXT STEPS

Within the OMG, several different initiatives have been geared towards supporting the embedded systems community, not least when it comes to language features for modeling real-time/embedded systems using UML. The real-time analysis and design working group was formed a few years ago, and agreed to a roadmap that included three different standard efforts:
– Modeling of schedulability, performance, and time characteristics
– Modeling of fault tolerance and quality of service characteristics
– Modeling of large-scale systems

The first of these have resulted in a UML profile that was completed last year [5]. The second effort was kicked off earlier this year when a request for proposal for another UML profile was issued [4]. Finally, the third area of the roadmap has been completely subsumed by the work on UML 2.0, and some of the results have been presented in this paper. Note that the profiles have been defined against UML 1.x, and cannot be used with UML 2.0 until they have been revised. Fortunately, the required changes are fairly minor.

5.1 Scheduling, Performance, and Time

The UML profile for Scheduling, Performance, and Time is usually referred to as the Real-Time UML profile, and focuses on the ability to capture information that is required for different kind of analysis. The general idea is that by annotating a model with information that is essential for a particular analysis technique, it is possible to integrate modeling tools with analysis tools dealing with for example schedulability and performance to make predictions about the behavior of the system under different conditions.

Examples of what the profile allows you to model include resources, scheduling policies, deadlines, execution times, and priorities. All of these are captured through stereotypes—the central UML extensibility mechanism—that may be attached to model elements of different kinds. All this flexibility comes at a price, though, since it is not always very easy to understand which properties must be used where to get useful results. The sheer amount of information that have to be put into a large model is also a hurdle that has to be overcome, for example through the use of automation in setting the values of the stereotypes. Used appropriately, however, the analysis tool can guide a user to create a model that satisfies specific requirements, such as the ability of a system to meet its deadlines or to have appropriate responses under different loads.

5.2   Testing

There is also a UML profile for testing in the works, where the initial submission has been delivered [7]. However, you should not expect results in this area until next year as it was decided to build the profile on UML 2.0, and which will allow the profile designers to exploit the new capabilities offered by sequence diagrams.

REFERENCES

1.   ITU-T, Z.100: Specification and Description Language (SDL), 2000
2.   ITU-T, Z.120: Message Sequence Charts (MSC), 2000
3.   JCP, UML/EJB™ Mapping Specification (JSR-26), http://www.jcp.org, 2001
4.   OMG, Action semantics for the UML, OMG ptc/02-01-09, 2002
5.   OMG, UML Profile for Modeling Quality of Service and Fault Tolerance Characteristics and Mechanisms RFP, OMG ad/02-01-07, 2002
6.   OMG, UML Profile for Schedulability, Performance, and Time Specification, OMG formal/01-09-67, 2001
7.   OMG, UML Profile for Testing Initial Submission, OMG ad/, 2002
8.   OMG, Unified Modeling Language Specification, version 1.4, formal/01-09-67, 2001
9.   Selic, B., Rumbaugh, J., Using UML for Modeling Complex Real-time Systems, Industrial white paper, Rational, 1998
10.   U2Partners, UML 2.0 Infrastructure Submission, U2Partners, OMG ad/02-06-01, 2002
11.   U2Partners, UML 2.0 Superstructure Submission, http://www.u2-partners.org, 2002

Chapter 10

EMBEDDED SYSTEM DESIGN USING UML AND PLATFORMS

Rong Chen[+], Marco Sgroi[+], Luciano Lavagno[++], Grant Martin[++], Alberto Sangiovanni-Vincentelli[+], Jan Rabaey[+]
[+]Univeristy of California at Berkeley, [++]Cadence Design Systems

Abstract: Important trends are emerging for the design of embedded systems: a) the use of highly programmable platforms, and b) the use of the Unified Modeling Language (UML) for embedded software development. We believe that the time has come to combine these two concepts into a unified embedded system development methodology. Although each concept is powerful in its own right, their combination magnifies the effective gains in productivity and implementation. This paper defines a UML profile, called UML Platform, and shows how it can be used to represent platforms. As an example, the Intercom platform designed at the Berkeley Wireless Research Center is presented to illustrate the approach.

Key words: UML, Platform-based Design, Embedded System, Quality of Service

1.   INTRODUCTION

Embedded System Design (ESD or just ES) is about the implementation of a set of functionalities satisfying a number of constraints ranging from performance to cost, emissions, power consumption and weight. Due to complexity increases, coupled with constantly evolving specifications, the interest in software-based implementations has risen to previously unseen levels because software is intrinsically flexible. Unlike traditional software design, embedded software (ESW) design must consider hard constraints such as reaction speed, memory footprint and power consumption of software because ESW is really an implementation choice of a functionality

E. Villar and J. Mermet (eds.), System Specification and Design Languages, 119–128.
© 2003 *Kluwer Academic Publishers.*

that can be also implemented as a hardware component, so that we cannot abstract away the hard characteristics of software. For this reason, we believe ESW needs to be linked 1) upwards in the abstraction levels to system functionality, 2) to the programmable platforms that support it, thus providing the much needed means to verify whether the constraints posed on ES are met.

UML is the emerging standard meta-notation (a family of related notations) used in the software world to define many aspects of object-oriented software systems [4]. Now UML is also capturing much attention in the ESW community as a possible solution for raising the level of abstraction to a point where productivity can be improved, errors can be easier to identify and correct, better documentation can be provided, and ESW designers can collaborate more effectively. An essential deficiency is that UML standardizes the syntax and semantics of diagrams, but not necessarily the detailed semantics of implementations of the functionality and structure of the diagrams in software. In [2], some requirements are identified that have to be satisfied to make UML a suitable development basis for embedded systems design, in terms of notation, semantics, refinement steps, and methodologies. The UML Platform profile described in this paper can be considered as a notation to satisfy some of the requirements of [2].

Platform-based design has emerged as one of the key development approaches for complex systems, including embedded systems in the last several years. In [1], an architecture platform is defined as a specific 'family' of micro-architectures, possibly oriented toward a particular class of problems that can be modified (extended or reduced) by the system developer. The choice of a platform is driven by cost and time-to-market considerations and is done after exploration of both the application and architecture design spaces. Furthermore, [1] defines an API platform, a software layer that abstracts computational resources and peripherals contained within the architecture platform, to hide unnecessary implementation details from embedded software developers. A platform can be described in terms of the type and quality of the services it offers to its users. Quality of Service (QoS) parameters, e.g. processing speed and I/O bandwidth, define platform performance and reliability and therefore are the essential distinguishing factors between platforms. The task of a designer is to find a platform that best meets the QoS requirements of applications [3].

When we combine UML with the platform-based design concept, we see, following the reasoning of [2], that it is necessary to have a way of describing those platforms in UML, i.e., a projection of the platform into the UML notation space. The definition of this projection is the purpose of this chapter that describes a "UML Platform" proposal.

2.   RELATED WORK

UML already has the capability to model the most relevant real-time system features, such as performance (using tagged attributes or OCL [10]), resources (using Component or Deployment Diagrams), and time (using classifiers and tagged attributes). However, in absence of a standard and unified modeling approach, the same embedded systems specification may be modeled in several different ways. Therefore, how to use UML for modeling real-time systems has become recently an active area of research and several proposals have been made.

The Real-Time UML profile [3] defines a unified framework to express the time, scheduling and performance aspects of a system. It is based on a set of notations that can be used by designers to build models of real-time systems annotated with relevant QoS parameters. The profile standardizes an extended UML notation to support the interoperability of modeling and analysis tools but touches little on platform representation. UML-RT [11] is a profile that extends UML with stereotyped active objects, called capsules, to represent system components. The internal behavior of a capsule is defined using statecharts; its interaction with other capsules takes place by means of protocols that define the sequence of signals exchanged through stereotyped objects called ports. The UML-RT profile defines a model with precise execution semantics; hence it is suitable to capture system behavior and support simulation or synthesis tools (e.g. Rose RT). HASoC [5] is a design methodology based on UML-RT notation. The design flow begins with a description of the system functionality initially given in use case diagrams and then in a UML-RT version properly extended to include annotations with mapping information. De Jong in [9] presents an approach that combines the informal notation of the UML Diagrams with the formal semantics of SDL. It consists of a flow from the initial specification phase to the deployment level that specifies the target architecture. The high-level system specification is specified using use case diagrams; the system components and their interactions are described using block diagrams and message sequence charts, respectively. Then the behavior of each module is specified using SDL that provides an executable and simulatable specification.

2.1   Our approach

In this chapter, we propose a new UML profile, called UML Platform, to model embedded system platforms. First, we introduce a subset of UML notation (new building blocks using stereotypes and tags) to represent specific platform concepts. Then, we show the main levels of abstraction for platforms and the most common types of relationships between platform components, and how to use appropriate UML diagrams along with aforementioned UML notations to model those platforms and relationships. Last, we explain how to represent platform QoS performance and do constraint budgeting.

2.2   Case study: the Intercom

The Intercom [7] is a single-cell wireless network supporting full-duplex voice communication among multiple mobile users in a small geographical area. Each remote can request one of the following services: subscription to enter active mode, unsubscription to return to idle mode, query of active users, conference with one or multiple remotes, and broadcast communication. The system specification also includes performance requirements on the transmission of voice samples such as latency, throughput, and power consumption. [7] defines a protocol stack that includes Application, Transport, MAC and Physical layers (see Figure 1).

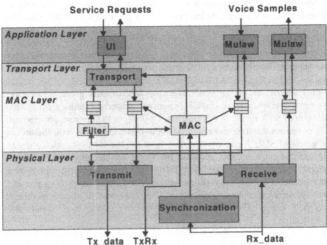

Figure 1. Intercom Protocol Stack

Its physical implementation is composed of a reconfigurable embedded processor (Tensilica Xtensa running the RTOS eCos), a memory subsystem, fixed and configurable logic and a silicon backplane (supporting Sonics OCP) that interconnects these components.

3. THE UML PLATFORM PROFILE

3.1 New stereotypes and tagged values

In this section, we introduce a set of new stereotypes and tagged values for the UML Platform profile. The list of stereotypes and tagged values is derived from the description of several platform examples and, hence, in our opinion it is sufficient to model most embedded platforms.

For each of the building blocks that are frequently used in modeling platform components such as processor, device driver, scheduler, table, buffer, memory, cache, etc., a stereotyped classifier is defined. A stereotyped classifier usually includes a set of attributes and methods that are specified only when the block is instantiated at modeling time. For example, the stereotyped class "cache" is associated with attributes such as "valid", "block index", "tag", and "data", methods such as "write through" or "write back", and QoS parameters such as "hit time" or "miss penalty".

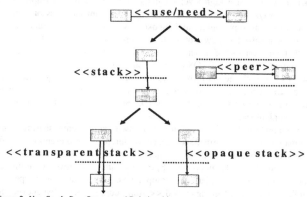

Figure 2. *Use, Stack, Peer* Stereotyped Relationship

The following stereotypes model common relationships between platform components. At the top of the hierarchy shown in Figure 2, the stereotype <<use>> represents a relationship in which an entity uses a service provided by a resource, while the stereotype <<need>> indicates when an entity needs a service from another entity, but the service is not currently available, i.e. not implemented. Thus it represents a request for future service enhancements.

<<stack>> and <<peer>> are refinements of the stereotype <<use>>. <<stack>> is used when the platform component providing the service and the one using it are at different levels of abstraction. We further specialize this stereotyped relationship into <<transparent stack>> and <<opaque stack>>. <<transparent stack>> models the case where the upper level component knows how the service is implemented within the lower level. So, it is possible for the upper entity to bypass the lower one in search of a service that is simpler, but more suited to the requirements. For example, normally a data transfer function interacts with a medium access control (MAC) function to transfer data, but if a faster transfer rate is desired the data transfer function may bypass the MAC function to directly call a device driver to access the network inter-connector. In such case, the data transfer function and the MAC function are related by a transparent stack relationship. <<opaque stack>> describes the case when the upper level component has no knowledge of how the service is implemented by the lower level component. Thus, the upper entity has to always rely on the lower one to provide the necessary service. For example, a platform service function written in a high-level language declared as an interrupt service routine always relies on an RTOS to save the microprocessor context, identify the interrupt source, and invoke it whenever the interrupt arrives. Due to the insufficient power of the high-level language, it cannot bypass the RTOS to run on top of a bare microprocessor. In this case, the platform service function and the RTOS form an opaque stack relationship. <<peer>> is used when both the platform component using and the one providing the service are at the same abstraction level. In general, peers can only exist within the same level platform, but stack can exist both within and across one level platform.

<<communicate>> is used to relate two components sharing some information. It can be further specialized by stereotypes representing specific models of computation, e.g. <<asynchronous>>, <<RPC synchronous>>, <<rendezvous>>, <<Kahn process>>, etc. <<coupling>> reveals the limited freedom in choosing platform components. There are two types of couplings: <<weak coupling>> and <<strong coupling>>. If whenever one entity is chosen, one from a certain group of entities must also be chosen in order to achieve some functionality, then we say a weak coupling exists between this entity and the group; if whenever one entity is chosen, exactly one other entity has also to be chosen in order to achieve some functionality, then a strong coupling exists between these two entities. Note that, although the <<coupling>> relationship can be also described in OCL [10], the stereotype form is preferred because it is more visible to users. Figure 3 shows an example of the use of <<weak coupling>> between an RTOS and a CPU. These two entities are coupled because when a CPU is used also an

RTOS must be used, and vice versa. This coupling is weak in both directions because there are several types of RTOS that can run on the Xtensa CPU and several CPUs that can support the eCOS RTOS. Finally, we call <<share>> the relationship among multiple entities that use services provided by the same resource (e.g. in Figure 3, tasks sender and receiver share the same CPU). In the presence of <<share>>, it is frequently necessary to deploy an allocation or arbitration scheme, such as a scheduler.

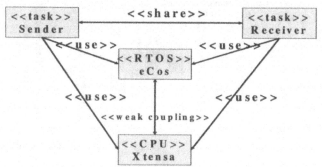

Figure 3. *Share, Coupling* Relationship

Tagged values (or called tags) are used to extend the properties of a UML building block with additional information. In the UML Platform profile, they are mainly used to specify QoS parameters, application requirements, as well as communication types, usage types, etc. Examples of relevant tags are: {throughput} for communication throughput, {delay} for time delay between request and response, {precision} for calculation precision, {power} for power consumption, {size} for memory size.

3.2   Platforms

We classify embedded system platforms into three abstraction levels: architecture (ARC), application programming interface (API), and application specific programmable (ASP) platforms (see Figure 4). The ARC Layer includes a specific family of micro-architectures (physical hardware), so that the UML deployment diagram is naturally chosen to represent the ARC platform. The API layer is a software abstraction layer wrapping ARC implementation details. API should be presented by showing what kinds of logical services (represented as interfaces) are provided and how they are grouped together. For example, it is important for users to know that preemption is supported by an RTOS, but not how this service is implemented inside the RTOS because users (either platform users or developers) rarely need to modify the RTOS itself. For such a purpose, RTOS, device-driver and network communication subsystem are treated as components, i.e., the physical packaging elements for logical services. In UML, their natural representation is the component diagram.

application domain-specific services (functions, user interfaces)			ASP platform
RTOS	network commu. subsystem	device driver	API platform
µP and memory	inter-connection	H W	I / o	ARC platform

high

abstraction level

low

Figure 4. Platforms at Different Levels

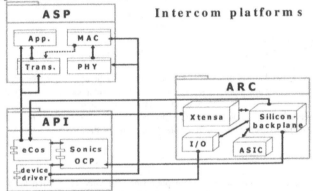

Figure 5. Intercom Represented by UML Platform

ASP is a platform, which makes a group of application domain-specific services directly available to users. For example, the function to set up a connection in the Intercom is such a domain-specific service. In addition to calling these existing services, users sometimes also need to modify or combine them, or even develop new services to meet certain requirements. Consequently, unlike API, here it becomes essential to show not only what functionality these services offer, but also how such services are supported by their internal structures, and how they relate to each other. In UML, the class diagram best represents such information.

Figure 5 shows how Intercom platforms are represented by UML Platform, where:

- ARC, API and ASP platforms are represented by deployment, component and class diagrams respectively;

- A <<transparent stack>> relationship exists within the ASP platform (such as the one indicated by the dotted line between Transport and MAC); an <<opaque stack>> relationship exists between ASP and API (such as the one indicated by the solid line between Transport and eCos), and between API and ARC (such as the one indicated by the solid line between eCos and Xtensa). This implies that Transport may bypass MAC in search of better-suited performance, but it can never bypass eCos;

- <<share>> is used twice: Application and Transport share eCos, while MAC and Physical share device driver.

3.3   Quality of Service (QoS) and Constraints

QoS parameters identify key performance properties and therefore allow classifying and comparing of different platforms. A set of QoS parameters that completely characterize the needs of any application is obviously impossible to define. However, it is possible to decompose QoS properties into just a few orthogonal dimensions (called QoS parameters). For example, for the ARC platform, such dimensions can be power consumption, memory size, processing capability, communication throughput, and for the API platform, task-handling number, task-waiting time, etc. The design constraints can be decomposed along the same dimensions, and this will conceivably enable some form of automatic matching and analysis between QoS properties and design constraints. In [6], the former are called *offered QoS* (the values are given by the platform providers), the latter *required QoS* (the constraints on these variables are specified by the platform users), and we adopt such terminologies here. In general, we model QoS parameters by annotating classifiers and relationships with tagged values. Constraints complement the functional specification of an ES with a set of formulae that declare the valid range for variables in the implementation. At the beginning of the design process, constraints are defined for the global system performance and then propagated (through *budgeting*) to lower levels of abstraction to bind the performances of local components as the design gets refined. The objective of this chapter is neither to define a methodology for constraint budgeting and verification nor to propose a new constraint specification formalism in addition to the ones in [8] and [10], but to provide guidelines on using the UML notation in the budgeting process: annotate diagrams with tags describing constraints and use the graph structure to show how the lower-level components are connected (either sequentially or concurrently) to drive the budgeting process. Assume a constraint ϕ is given on the minimum throughput of a node, and this node is refined into multiple components. If two components are composed in sequence, ϕ is propagated to both, because throughput is non-additive for series composition. Instead, if two components are composed in parallel, two constraints, $\phi1$ and $\phi2$, can be derived, provided their sum does not exceed the original constraint ϕ.

4.   CONCLUSIONS

In this paper we have presented a new UML Profile, called UML Platform. We have introduced new building blocks to represent specific platform concepts; selected proper UML diagrams and notations to model platforms at different abstraction levels including QoS performance and constraints. As future work we will develop a full design methodology and tool set based on UML Platform.

5.   REFERENCES

[1] A. Sangiovanni-Vincentelli, Defining Platform-based Design, EEDesign, Feb 2002.
[2] G. Martin, L. Lavagno, J. Louis-Guerin, Embedded UML: a merger of real-time UML and co-design, Proceedings of CODES 2001, Copenhagen, Apr. '01, p.23-28.
[3] Bran Selic, A Generic Framework for Modeling Resources with UML, IEEE Computer, June 2000, pp.64-69
[4] J. Rumbaugh, I. Jacobson, and G. Booch, The Unified Modeling Language User Guide, Addison-Wesley, 1998
[5] P. N. Green, M. D. Edwards, The modeling of Embedded Systems Using HASoC, Proceedings of DATE 02.
[6] ARTiSAN Software Tools, Inc. et al., Response to the OMG RFP for Schedulability, Performance, and Time, OMG document number: ad/2001-06-14, June, 2001.
[7] J. da Silva Jr., M. Sgroi, F. De Bernardinis, S.F Li, A. Sangiovanni-Vincentelli and J. Rabaey, Wireless Protocols Design: Challenges and Opportunities. Proceedings of the 8th IEEE International Workshop on Hardware/Software Codesign, CODES '00, S.Diego, CA, USA, May 2000.
[8] Rosetta, www.sldl.org
[9] G. de Jong, A UML-Based Design Methodology for Real-Time and Embedded Systems, Proceedings of DATE 02.
[10]   J. Warmer, A. Kleppe, The Object Constraint Language: Precise Modeling with UML, Object Technology Series, Addison-Wesley, 1999.
[11]   Selic, J. Rumbaugh, Using UML for Modeling Complex Real-Time Systems, White paper, Rational (Object Time), March 1998.

Chapter 11

A DESIGN METHODOLOGY FOR THE DEVELOPMENT OF A COMPLEX SYSTEM-ON-CHIP USING UML AND EXECUTABLE SYSTEM MODELS

Marc Pauwels {1}, Yves Vanderperren {1}, Geert Sonck {1}, Paul van Oostende {1}, Wim Dehaene {2}, Trevor Moore {3}
{1} STMicroelectronics Belgium (previously with Alcatel Microelectronics), {2} Katholieke Universiteit Leuven, {3} Holistic Systems Engineering Belgium

Abstract: This paper describes aspects of the process and methodologies used in the development of a complex system-on-chip. The open-standard modelling language SystemC played a key role in supporting the technical work, from early architectural modelling, through a defined refinement process to detailed cycle-accurate modelling elements, which enabled early co-simulation and validation work. In addition to SystemC, significant use was made of the Unified Modelling Language, and process and methodology associated with it, to provide visual, structured models and documentation of the architecture and design as it developed.

1. INTRODUCTION

The challenges presented to the design community by the ever-greater potential of System-On-Chip technology are well documented [1][2]. SystemC [3] provides the designer with an executable language for specifying and validating designs at multiple levels of abstraction. We decided to adopt SystemC as an integral part of the design process for a recent System-On-Chip (SoC) development. The product under development was a wireless LAN chipset supporting more than one networking standard sharing very similar physical layer requirements [4][5]. To ease integration of the devices into larger systems, significant higher-level protocol complexity was required - in addition to the already complex transmit and receive signal processing in the physical layer. To efficiently and flexibly

E. Villar and J. Mermet (eds.), System Specification and Design Languages, 129–141.
© 2003 *Kluwer Academic Publishers.*

address the implementation of the signal processing functions of the physical layer, an architectural approach was adopted which made use of a number of custom-designed, microprogrammable engines in combination with general-purpose processing cores. The scale of the project, both in terms of complexity and the degree of architectural novelty, demanded that the project team consider adopting new approaches in order to manage the risks involved in the development. In addition to SystemC, significant use was made of the Unified Modelling Language [6], and process and methodology associated with it, to provide visual, structured models and documentation of the architecture and design as it developed.

2.   AN OVERVIEW OF THE METHODOLOGY

The presented SoC Methodology was a fusion of some of the best ideas from the digital hardware and software engineering worlds.

The engineering team had significant previous experience of systems development using UML tools and associated methods. Other members of the team had long experience of digital hardware design using VHDL and recognised the importance of adopting the modelling and abstraction techniques offered by SystemC in order to tackle ever more complex designs. Key central ideas from this combined experience formed the foundation of our approach:

- **Iterative Development** – based on the ideas originally promoted by Boehm's spiral model [7] and central to many modern software development processes [8]. The project is structured around a number of analysis, design, implement and test cycles each of which is explicitly planned to address and manage a key set of risks. The project team maintains the risk list as the project unfolds and knowledge and understanding of the issues develops. Each iteration is targeted to resolve or mitigate a number of these.
- **Use Case Driven Architecture** – the functional requirements of the *system* (not just the hardware or the software components) are analysed in terms of Jacobson Use Cases [9]. This approach has been found to be very effective in several important ways:-
    - ◊ It enables a structured statement of the system requirements – resulting in improved understanding across the team – and requirements validation with the customer/market representative
    - ◊ It drives the selection and assessment of the product architecture

     ◊   It provides a foundation for the estimation, planning and management of project iterations, as well as an excellent basis for system verification planning

- **Proactive Requirements Management** – in a fluid requirements environment, it is vital to maintain an accurate record of the project requirements. The tracing of functional requirements to Use Cases provides a powerful complexity-management technique and ensures straightforward review and approval of system verification specifications.

- **Executable System Models and Model-Centric Development** – the iterative design philosophy demands the ability to approach the design of the system as a series of explorations and refinements, each of which delivers answers to key questions and concerns about the design. The ability to model the system and its environment at different levels of abstraction is vital to support such an approach.

The presented process did not attempt to modify the existing processes for detailed design and implementation. Rather it provides a comprehensive approach to the management of the transition from customer requirements through architectural design to subsystem specification (the entry point to detailed design), which is vital to control technical risks in a complex, multi-site development project.

3. REQUIREMENTS CAPTURE AND USE CASE ANALYSIS

3.1 The Vision Document

One of the first activities was the drafting and review of the so-called Vision Document. Though the Rational Unified Process [8] advocates such a document in a software engineering context, the value of the document is no less for a system development. The Vision Document is the highest level requirements document in the project and provides:

- A statement of the business opportunity and anticipated positioning of the product in the marketplace and the competitive situation
- Identification of all anticipated users and stakeholders[1]
- Identification of the success criteria and needs for each user and stakeholder

---

[1]   A stakeholder has an interest in the project success, but may not use the system directly

- Identification of the key product Features and Use Cases. Features are the high-level capabilities of the system that are necessary to deliver benefits to the users.
- Lists design constraints and other system-level non-functional requirements (e.g. performance requirements), etc[2].

The process of agreeing the contents of the Vision Document can prove cathartic. It is not uncommon for several different views of the project requirements to co-exist within the project and early alignment around a clearly expressed set of goals is an important step.

3.2   Requirements Capture

Detailed requirements capture takes place from a number of sources. In the case of the wireless LAN project, the primary source documents were the published networking standards to which the product needed to comply. The requirement database was built up and structured according to the Features and Use Cases identified in the Vision document. Additional project requirements could be identified by consideration of product features and user needs which the standards do not address. The resulting database allowed a number of requirements views to be extracted, from a comprehensive listing of all project requirements structured by feature, to a high-level feature list useful for discussing iteration planning.

3.3   Use Case Analysis

The Use Case driven approach to software engineering was formalised by Jacobson [9], and has much in common with earlier ideas of system event-response analysis [10]. Use Cases would normally be documented in textual form, augmented by sequence diagrams showing the interactions between the system and its environment. The UML Use Case Diagram notation assists in visualising the relationships between Use Cases and the system actors. It is important at this stage to avoid any assumptions about the internal design of the system. This approach to the analysis of functional requirements is in no way restricted to software-dominated systems, but provides a very general method for structuring functional requirements of any system, including a complete System-On-Chip.

---

[2]   Key measures of effectiveness can be quantified here – non-functional parameters that are essential for product success (e.g. power consumption, cost) can be identified to focus design attention.

4. MODELLING STRATEGIES

Modelling in various forms enables an iterative approach to systems specification. The primary deliverables of the modelling activity are proven subsystem specification and test environments as input to detailed design. Effort invested in creating the models is repaid by faster subsystem design and integration cycle times because the specifications contain less uncertainties and errors and are more clearly communicated.

4.1 Algorithmic Modelling

The requirements for the physical layer of the system were heavily dominated by the need to specify signal-processing algorithms. Many required significant research and analysis. For such work the team made full use of an optimised toolset (Matlab) to develop and verify the algorithms. This work followed conventional lines and is not described further.

Once verified, the Matlab model became the ultimate reference for the performance of the key signal processing algorithms (refer to section 5.3).

4.2 SystemC Modelling

4.2.1 Model Development

SystemC was used to develop an executable model of the whole system architecture to gain confidence in the decisions, which had been made. Example concerns were:

- Would key sections be able to process data fast enough operating at critical clock rates?
- Would the chosen architecture implement the designed algorithms with sufficient accuracy to reproduce the performance obtained in the Matlab environment?
- The need to specify control and configuration interfaces in detail to enable control plane software to be designed.

The initial behavioural requirements are provided by use cases and, for signal-processing oriented subsystems, by mathematical models in Matlab. The first modelling phase represents each subsystem block at the functional level. Using this approach a model of the system can be constructed using a high level of abstraction – with the benefits of speed of construction and execution - to gain confidence and insight on the high level subsystem interactions. Adding information on throughput and latency targets to the model allowed a broad timing overview to be constructed. These models could be used as the baseline specification for detailed design in their own right. The SystemC model can be further refined to provide cycle accurate simulation of communication busses or taken as far as RTL-equivalent cycle accurate simulation of logic blocks. At any stage the complete design can be simulated using a mixture of these different levels, providing flexibility to explore the design at different levels as appropriate to the implementation risk. Co-simulation of SystemC and VHDL may be carried out in order to verify project-designed VHDL in the system environment, or in order to incorporate 3rd party IP delivered in the form of VHDL.

In general, of course, as the level of abstraction is lowered the accuracy and confidence of the results goes up, but so does the time required for simulation. A SystemC model provides very useful flexibility to construct verification environments, which effectively trade off these two concerns.

4.2.2   Summary of Experience

The following conclusions were drawn from the modelling activity:
- SystemC is just a language, a sound methodology and process is necessary to support its application in a project of significant size
- Many disciplines familiar to software engineering must be applied to the development of the model – for example
  ◊ Clear coding standards
  ◊ Good control of overall architecture and interface definition between developers to avoid model integration problems
- Careful attention to coaching and mentoring eases the adoption of the new techniques and helps to build consistent approaches to design

It is not unimportant to note that the activity of building and testing the SystemC model serves as an excellent catalyst for cross-functional communication. The project not only gave the design team members deep familiarity with the content of the model – making the transition to detailed design much smoother – but also brought together representatives from different disciplines with the clear and tangible objective of modelling the system – encouraging cross-fertilisation of ideas and techniques.

4.3   UML Modelling of Architecture

UML was applied early in the project to the analysis of requirements and to the specification of higher layer software, using modelling and process guidelines based on the RUP [8]. The details of the software modelling approach are outside the scope of this paper.

An architectural model was developed using a UML tool with the following objectives:

1. To provide a common environment for the specification of high level architectural decisions across the disciplines
2. To specify the black-box requirements for each major subsystem, which could be implemented in a variety of ways – from software on a general-purpose core, through assembler code on a custom processor to dedicated hardware.
3. To specify in detail the requirements for inter-subsystem interfaces
4. To provide a documented framework for the structure of the SystemC model – each major subsystem mapped to a sc_module.

A modelling exercise of this scope required a robust structure to manage its development and to allow useful abstractions to be maintained. The following subsections discuss the most important principles established.

4.3.1   Use of UML Stereotypes

The UML provides extension mechanisms to apply the language to specialised domains. Stereotypes to represent important concepts of structure (subsystem, channel, interface) were defined and supported the layering of the model described in the following section. Stereotypes representing SystemC concepts (<<sc_module>>, <<sc_port>>) allowed an explicit mapping between the UML and SystemC models.

4.3.2   Model Layering

The most important abstraction adopted in the architectural modelling was the separation of high level domain-relevant behavioural modelling from the details of subsystem interfacing (Figure 1).

At the higher level, subsystems interact with each other using "logical" interfaces provided by the communication channels which connect them. This interaction occurs at a high level of abstraction and focuses on how the subsystems work together to meet the requirements – for example a typical logical primitive might be SendPDU(NetworkAddress, PduData).

The lower level of the model deals with the detailed design of the communications interfaces (physical medium details, e.g. a bus) and what further processing is required in order to support the high level "logical" interface over the chosen medium. A typical design level interface primitive might, for example, be WriteByte(RamAddress, ByteData).

Figure 2 shows a typical structural diagram of the logical view of part of the system and Figure 3 shows a design level view of the same channel.

Figure 1. Model layering separating behaviour from communication

Figure 2. Logical view of subsystems and channels

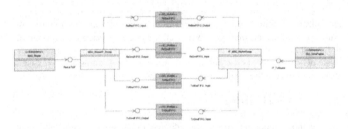

Figure 3. Design view of channel showing implementation entities

This technique allowed clean separation of detailed issues of software/hardware interaction to be considered within a framework, which clearly captured the overall intent of the interface. The explicit identification in the model of hardware entities (e.g. dual port RAM) allowed associated design information (such as e.g. memory maps) to be directly linked to the model browser. Thus the model also acts as a powerful project "organiser".

The model is developed by documenting Use Case Realisations using UML sequence diagrams. These realisations illustrate how the architectural elements interact in order to fulfil a given Use Case scenario. This is an extremely useful, although simple, method of illustrating dynamic interactions between elements of the model and allows details of interactions between software subsystems and hardware entities to be analysed.

4.3.3   Summary of Experience

The described application of UML provided significant benefits, a.o.:
- A common, structured environment for the documentation of system requirements and design information
- Enhanced inter-disciplinary communication – e.g. sequence diagrams showing complex software-hardware interactions reduced misunderstandings between the disciplines earlier and more directly than would be expected with conventional specification approaches
- Usefully separates abstraction levels to aid understanding

Of course, there have to be some issues – chief amongst these are:-
- The architectural model is based on an *a priori* decision about the partitioning of functionality in the system. If the architecture is modified, the model needs to be manually brought into line (but then so would a textual document).
- There is no direct link between the UML model and the SystemC model. Many UML tools provide a "round-trip" capability between the code and the model to ensure the two views remain synchronised and it would be very useful to explore the possibility of exploiting such capability for a SystemC model.
- Engineers without a UML or OOA background need to be introduced to the modelling concepts, process and notations. On the one hand this is a steep learning curve, on the other – if digital design engineers are going to move into the world of object-oriented executable modelling then a grounding in current best-practices from the software industry can hardly be a bad thing.

5.   ITERATIVE DEVELOPMENT, MODEL REFINEMENT AND VERIFICATION

5.1   Iterative and Model-Based Design Methodology

It is generally accepted in software engineering circles that one should strive to keep development iterations short. Too lengthy an iteration risks investing too much of the project resources, which makes it difficult to backtrack if better approaches become apparent in the light of the iteration experience. It is also important that the result of the iteration delivers output of real value in illuminating the way ahead for the team.

The wireless LAN project iterations are shown in table *11-1*. The iteration plan may seem sluggish to software developers used to significantly more rapid cycles, but independent teams within the project were of course free to execute their own, more frequent iterations within the scope of a single project iteration.

Table 11-1. Project iterations summary

	S	It0	It1	It2	It3
	Feasibility, Requirements and Algorithms	Architecture and High Level Modelling	Detailed Design and Cosimulation	Hardware Prototype (FPGA)	Silicon
Systems	Capture reqts; Agree Vision Doc; Create UC model; Develop key algo's	Specify Architecture; Create HL SystemC / UML model; Demonstrate architecture can meet project requirements	Cosimulation of SystemC, VHDL and SW	Cosimulation of SW on target and HW on FPGA; Conduct system V&V	Silicon Verification; Product Qualification or Approvals
Hardware		Involvement with SystemC specification; Early trials of SystemC to VHDL mapping	Detailed Design of VHDL	Port design to FPGA	Backend design and silicon fab
Software		Specify SW Arch; Create Host-Based SW framework and basic functionality	Build further SW functionality on It0 framework; Port to target	Build further functionality on It0/It1 framework	Build further SW functionality

Section 4.2 described the multiple levels of modelling which SystemC supports. During Iteration 0, for example, sections of the design that were considered high risk were modelled at CA level and co-simulated with TF-level blocks. During Iterations 1 and 2 a number of cosimulation scenarios were planned, involving the use of an Instruction Set Simulator to support co-verification of VHDL, SystemC and target software. Limited space prevents the presentation of further detail in this paper.

5.2   Workflows for Detailed Design

The modelling and simulation activity provides a verified, executable specification as input to detailed design of software, firmware and hardware. The intent is that the model should be functionally correct. The design challenges to implement target software, which is performant and meets the resource constraints of the target and detailed synthesisable RTL hardware definitions remain to be solved during the detailed design phase. The philosophy adopted during the project was that, at least at this stage of maturity of methods and tools, the translation from specification to detailed implementation must remain a manual design activity using tools and techniques appropriate to the discipline.

5.3   System Verification & Validation

Most early test activity deals with ensuring that the early architectural and analysis work correctly delivers the specified functionality and performance. One of the strengths of the presented approach is the support it provides for the development of system verification environments (testbenches) in tandem with the design of the system itself. These testbenches are maintained and reused throughout the project for system, subsystem and component testing, and may even be incorporated into product code to provide Built In Self Test functionality to the final product.

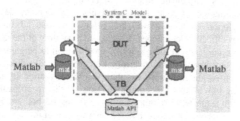

Figure 4. Matlab as test I/O for SystemC

5.3.1   SystemC Verification from Matlab

The Matlab model is the algorithmic reference for the system. In order to compare the performance of the SystemC model against the original algorithms, a bridge between the Matlab tool and the SystemC environment was required. This was achieved by using Matlab APIs from within SystemC (Figure 4). In this manner, the input stimuli for the SystemC model were extracted from Matlab and output results of the SystemC model written to a Matlab file for comparison and analysis in the Matlab environment.

To enable analysis of the effect on performance of operating in fixed or floating point, a new data type called *fx_double* was used in the SystemC model which allows one executable to be used for analysing both floating and fixed point results [11].

5.3.2   System Validation

It is often the case with complex wireless systems that final proof of the system's capability to meet customer expectations must wait until early prototypes can be tested in realistic environments. In our case, the FPGA-based prototype provides a pre-silicon platform to gain confidence in the anticipated performance of the system and its algorithms in the real world. It is important that controlled test environments (eg radio channel simulators) are considered to allow the performance to be explored.

6.   CONCLUSIONS

The project team recognised the need to adopt untried practices in order to manage successfully the development of a complex design.

The combination of tools and techniques used for the development proved to support each other well and offer many real advantages over traditional specification techniques. Executable modelling of the architecture allowed real confidence in the quality of specification to be gained early in the project – many specification errors were undoubtedly discovered and rectified much earlier in the process than would conventionally have been expected. On top of this the communication of the specification to the detailed design teams was smoother and less error-prone, both because of the formal nature of the specification and the involvement of the designers in the modelling process.

Of course much work remains to be done to improve and streamline the presented foundations. Ultimately any development approach must satisfy a cost-benefit equation to the satisfaction of the organisation applying it. In the opinion of the authors the additional costs of the approach described in this paper are more than outweighed by the benefits of:

- **Improved product quality** – both in terms of meeting initial requirements and robust and flexible architectures capable of absorbing inevitable requirements changes
- **Improved scheduling accuracy** – whilst it may be over-optimistic at this stage to claim significant benefits in overall time to market for this approach, the team would certainly claim that it provides a much sounder basis for confident prediction of product release date than the traditional waterfall development
- **Improved inter-disciplinary cooperation** – fostered by the activity of system level modelling and yielding real dividends in cross-fertilisation of ideas and better communications
- **Improved team confidence** – the iterative approach provides many psychological benefits to the team members. Early and regular demonstrations of progress not only gives management important confidence, but provides satisfaction to the engineering staff.

ACKNOWLEDGEMENTS

The authors would like to thank Jacques Wenin for his support and far-sighted appreciation of the value of process improvement.

7. REFERENCES

[1] On Nanoscale Integration and Gigascale Complexity in the Post .Com World, Hugo de Man, DATE Conf Proceedings, 2002

[2] Global Responsibilities in SoC Design, Taylor Scanlon, DATE Conf Proceedings, 2002

[3] SystemC User Guide (v2.0), Open SystemC Initiative, www.systemc.org

[4] Broadband Radio Access Networks (BRAN); HIPERLAN Type 2; System overview, ETSI TR 101 683 (v1.1.2)

[5] Wireless LAN Medium Access Control (MAC) and Physical Layer (PHY) Specifications , ANSI/IEEE Std 802.11, 1999 Edition

[6] UML Standard v1.4, Object Management Group, www.omg.org

[7] A Spiral Model of Software Development and Enhancement, B. W. Boehm, IEEE Computer Vol 21, No5 1988

[8] Rational Unified Process, Rational Software Corporation, www.rational.com

[9] *Object-Oriented Software Engineering: A Use Case Driven Approach,* I. Jacobson, M. Christerson, P. Jonsson and G. Övergaard, Addison-Wesley, Wokingham (England), 1992

[10] *Structured Development for Real-Time Systems,* P. Ward and S. Mellor, Yourdon Press, 1985

[11] A Method for the Development of Combined Floating- and Fixed-Point SystemC Models, Y. Vanderperren, W. Dehaene and M. Pauwels, Conf Proceedings FDL '02

Chapter 12

USING THE SHE METHOD FOR UML-BASED PERFORMANCE MODELING

B.D. Theelen, P.H.A. van der Putten and J.P.M. Voeten
Information and Communication Systems Group, Faculty of Electrical Engineering, Eindhoven University of Technology. P.O. Box 513, 5600 MB Eindhoven, The Netherlands. E-mail: B.D.Theelen@tue.nl

Abstract: The design of complex real-time distributed hardware/software systems commonly involves evaluating the performance of several design alternatives. Early in the design process, it is therefore desirable that design methods support constructing abstract models for the purpose of analysis. Recent extensions to the Unified Modeling Language (UML) that enable specifying schedulability, performance and time provide a means to start developing such models directly after defining the concepts and requirements of a system. However, UML hampers the evaluation of performance properties because this requires constructing executable models with a modeling language that supports application of mathematical analysis techniques. In this paper, we present how the Software/Hardware Engineering (SHE) method can be used for the performance modeling of real-time distributed hardware/software systems. Starting from a UML specification, SHE enables constructing formal executable models based on the expressive modeling language POOSL (Parallel Object-Oriented Specification Language).

Keywords: Formal Semantics, Performance Modeling, Parallel Object-Oriented Specification Language (POOSL), System-Level Design, and Unified Modeling Language (UML).

1. INTRODUCTION

Designing complex real-time distributed hardware/software systems entails considering different options for realizing the demanded functionality. In the early phases of the design process, deciding for a certain design alternative may have a deep impact on the final performance of the system. To evaluate the performance of design alternatives before actually

E. Villar and J. Mermet (eds.), System Specification and Design Languages, 143–160.
© 2003 *Kluwer Academic Publishers.*

implementing the system in hardware and software, system-level design methods can be applied. System-level design methods structure the early phases of the design process by assisting the designer in developing abstract models of the system for analyzing its properties [2]. To support performance modeling, system-level design methods provide heuristics for applying certain modeling languages and performance analysis techniques. Additionally, user-friendly computer support is often supplied in the form of tools that enable efficient application of the analysis techniques by executing the constructed models.

An example of a modeling language is the Unified Modeling Language (UML) [14], which provides a set of graphical notations for specifying the functionality of a system. UML is often used to initiate and stimulate discussions on the system's concepts and for the documentation of these concepts. In principle, UML specifications are not executable, which hampers analysis of especially large real-life industrial systems. UML tools, however, often allow supplementing UML diagrams with executable code using a host language such as C++. These tools provide additional support for handling important aspects like concurrency and time because the semantics of UML does not describe how to manage them. Recent extensions to UML provide a standardized way for denoting concurrency and time aspects of a system [16, 4]. Although this real-time version of UML additionally allows capturing performance requirements and QoS characteristics, application of mathematical analysis techniques remains complicated due to the difficulty of relating such formal techniques to the informal UML diagrams.

To assist the designer in analyzing properties of a system during the early stages of the design process, the Software/Hardware Engineering (SHE) [13, 3] method was introduced. SHE provides heuristics for developing UML models that can be transformed into executable models specified in the Parallel Object-Oriented Specification Language (POOSL) [13, 1]. POOSL is an expressive modeling language with a formal (i.e., mathematically defined) semantics, intended for analyzing complex real-time distributed hardware/software systems. Based on the formal semantics, POOSL allows mathematical reasoning about the performance of a system. It has proven to be successful for modeling and analyzing real-life industrial systems. In [18] for instance, POOSL was applied for evaluating the performance of design alternatives for an industrial Internet router. This router basically concerns an input/output buffered switch system that is protected by a flow control mechanism. In this paper, we elaborate on the SHE method and discuss the UML profile it defines for developing executable performance models with POOSL. The Internet router of [18] is used as an example to illustrate several aspects of applying the SHE method.

The remainder of this paper is structured as follows. The next section presents the system-level design flow on which performance modeling with SHE is based. In section 3, the main characteristics of the UML profile for SHE are discussed, whereas the evaluation of performance properties is examined in section 4. In section 5, an overview is given of the tools provided by SHE. Section 6 briefly compares the SHE method with other methods for UML-based performance modeling. Conclusions and directions for future work are summarized in section 7.

2. SYSTEM-LEVEL DESIGN FLOW

Using the SHE method for performance modeling is based on the (idealized) system-level design flow presented in Figure 1. System-level design starts with discussions and brainstorm-sessions on *concepts* for realizing the demanded functionality. In addition, a set of performance *requirements* that are to be satisfied by the final implementation is defined. Design experience and design decisions taken in the past may affect these concepts and requirements. Since discussions and brainstorm-sessions have a rather unstructured and undocumented character, a stage of *formulation* is involved. SHE supports the formulation stage with Object-Oriented Analysis (OOA) techniques and utilizes UML diagrams to formulate the concepts in a *concept model*. The requirements are formulated as *questions*, which actually concerns determining the design issues that need to be assessed. The result of the formulation stage is a structured (but informal) description of the concepts and requirements using schematic pictures and plain texts, composing the deliverable at milestone A (see Figure 1).

After formulating the concepts and requirements, a stage of *formalization* is involved because the concept model is not executable. Formalizing the concept model concerns developing an *executable model* with the formal modeling language POOSL. The UML profile used in the formulation stage smoothens the application of POOSL for developing such models. *Validation* ensures that the executable model represents the (informal) concept model in a way that is adequate for answering the questions of interest. This is important considering the need for abstraction from implementation details that are either irrelevant for answering the questions or yet unknown in this early phase of the design. In parallel with formalizing the concept model, POOSL is used to formalize the questions into *formal properties*. These concern mathematical formulae specifying relevant performance metrics. As indicated in Figure 1, the stage of formalizing the concept model and questions is completed with the deliverable at milestone B. It documents the validated executable model and formal properties.

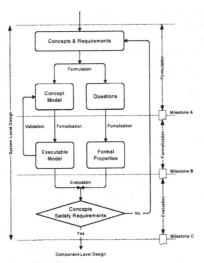

Figure 1. Formulation, Formalization and Evaluation in System-level Design.

In the final stage, the formal properties are *evaluated* against the executable model. Having a formal semantics, POOSL allows application of mathematical analysis techniques to evaluate performance metrics either by analytic computation or empiric simulation. Founded on the evaluation results, it can be concluded whether the concepts satisfy the requirements. If some requirement is not satisfied, a *design decision* can be taken based on the obtained evaluation results. Such design decisions may change the concepts for realizing the demanded functionality, the requirements or both. In these cases, the formulation, formalization and evaluation stages must be repeated* and the deliverable at milestone C contains the evaluation results and conclusions to found the design decision. However, if all requirements are satisfied, the deliverable at milestone C documents the evaluation results and conclusions as initial requirements for the individual components of the

---

* When concurrently applying the three stages for several design alternatives, which are all expected to satisfy the requirements, an 'optimal' design alternative might be chosen based on the evaluation results. Sometimes, a parameterized model can be used to evaluate several design alternatives (See for example the Internet router of [18]). However, if none of the design alternatives satisfies the requirements, the three stages really have to be repeated to devise new concepts for realizing the demanded functionality.

system. As shown in Figure 1, detailed component-level design can then be initiated.

3. UML PROFILE FOR SHE

To support developing executable models, SHE uses a UML profile in the formulation phase that expresses important characteristics of POOSL. In this section, we elaborate on the main aspects of this UML profile.

3.1 Objects and Classes

Similar to the suggestion in [4], SHE distinguishes two types of resources for complex real-time distributed hardware/software systems. Active resources concern components of the system that may take the initiative to perform specific functions without involving any other resource.

Active resources can often be arranged according to a hierarchical structure. For instance, a system may incorporate several complex components, which all consist of some configuration of basic components. Passive resources, on the other hand, present information in a system that is generated, exchanged, interpreted, modified and consumed by active resources. Figure 2 depicts the three stereotype classes that SHE uses to specify different resource types.

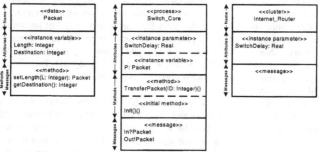

Figure 2. Data, process, and cluster classes.

SHE offers *data objects*, which are instances of *data classes*, to model passive resources of hardware/software systems. Their attributes are specified as *instance variables*. The sequential behavior of data objects is described with *(data) methods*. Data objects may receive messages, which leads to the atomic execution of equally named methods. Upon completion of executing a method, the result of a calculation or the data object itself is returned. We remark that data objects are comparable to objects in traditional object-oriented languages such as Smalltalk.

To specify the real-time behavior of (non-composite) active resources, SHE provides process objects or processes. Processes are instances of process classes. Their attributes are specified either as instance variables or as instance parameters. Instance parameters allow parameterizing the behavior of a process at instantiation. We remark that such parameterization is especially useful for analyzing several design alternatives with a single model. Processes perform their behavior independently from each other, in an autonomous (asynchronous) concurrent way. The behavior of processes is described with sequentially or concurrently executed (process) methods. Methods may be called with parameters and may return results through return variables. Methods without return results can be called tail-recursively, which offers an intuitive way to model infinitely repeating behavior. The start behavior of a process is defined with a unique initial method.

Processes and clusters (see below) are statically interconnected by *channels*. Channels model any possibility to exchange information between components as for example specified in collaboration or sequence diagrams. Channels interconnect processes and clusters independent from communication protocols and may reflect any topology [13]. Processes can communicate (copies of) their encapsulated data objects by (synchronously) passing messages over channels through *ports*. Next to the use of standard UML message types, SHE defines some additional message stereotypes. As POOSL only incorporates a statement for synchronous message passing, a set of modeling patterns is provided for formalizing the other types of messages. Notice that messages exchanged between processes are not directly related to their methods (as is the case for data objects). Therefore, a separate compartment in process classes specifies the messages it can send and receive, see Figure 2.

To model composite active resources, SHE provides *clusters*. Clusters are instances of *cluster classes* and group a set of processes and clusters (of other cluster classes). The need for using a cluster emerges from aggregation relationships between classes that represent active resources. Parameterized behavior of processes and clusters incorporated in a cluster can be initialized via the *instantiation parameters* of that cluster. The behavior of clusters is defined by the parallel composition of the incorporated processes (possibly interconnected by channels) and therefore does not extent the behavior of these processes. Consequently, cluster classes do not include a method compartment. Messages of incorporated processes (or clusters) that may pass the cluster's boundary through ports are specified in the message compartment.

Focusing on class diagrams, the identification of specialization/generalization relationships commonly results in using the inheritance and method overriding capabilities of POOSL. We remark that specialization/generalization relationships are only possible between passive resources on one hand and between active resources on the other hand. Any relationship in class diagrams other than specialization/generalization and aggregation has to be expressed by the behavior of the involved objects. Another important aspect of using SHE is the need for modeling the environment as specified with actors in use case diagrams. Being active resources, actors are modeled with processes and clusters. The classes representing actors therefore emerge in class diagrams as well.

3.2   Architectural Structure

Processes, channels and clusters provide the means for specifying the hierarchical structure, topology or implementation boundaries of real-time distributed hardware/software systems [13]. These aspects are commonly reflected in deployment, component or collaboration diagrams. Although supporting these types of specifications, using SHE involves deriving so-called *instance structure diagrams*, which formally define the architecture of models. Instance structure diagrams graphically represent the static structure of processes and clusters in a way that is independent from scenarios [13].

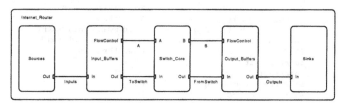

Figure 3. Instance structure diagram of the cluster class named Internet_Router.

Figure 3 depicts an instance structure diagram for the Internet router example of [18], where similar physical resources were modeled using a single process.

Channels are introduced when processes or clusters communicate in some scenario that is reflected in for example collaboration or sequence diagrams. Such channels are labeled with a name and are connected to the ports of the involved processes and clusters, which are indicated with a small rectangle at their edges. In instance structure diagrams, no information on the order of messages nor on the type of message is represented. We remark that one channel may be used to indicate the possibility of communicating different message types between the same components.

3.3   Behavior of Processes

To formalize the behavior of active resources, which is often specified using state or activity diagrams, POOSL offers the statements indicated in table 1.

Method abstraction and the statements for sequential and parallel composition provide ample means for formalizing the behavioral hierarchy that is sometimes expressed in state and activity diagrams. The statement for parallel composition denotes the purely interleaved (i.e., non-communicating) execution of statements S_1 through S_n. With this statement, POOSL offers (asynchronous) concurrency *within* processes next to the usual concurrency *among* processes. The concurrent activities specified with S_1 through S_n may share the data objects assigned to the instance parameters and instance variables of the involved process. Operations on data objects are atomic (i.e., indivisible), which solves mutual exclusion problems in a natural way.

Table 1. Statements for specifying the behavior of processes.

Statement S	Description
$m(E_1, \ldots, E_i)(v_1, \ldots, v_j)$	Method Abstraction
par S_1 **and** S_2 **and** ... **and** S_n **rap**	Parallel Composition
$S_1; \ldots; S_n$	Sequential Composition
$E_p!m(E_1, \ldots, E_i)\{E_{at}\}$	Message Send
$E_p?m(v_1, \ldots, v_i \mid E_{rc})\{E_{at}\}$	(Conditional) Message Receive
sel S_1 **or** S_2 **or** ... **or** S_n **les**	Non-deterministic Selection
$[E]\, S$	Guarded Execution
abort S_1 **with** S_2	Abort
Interrupt S_1 **with** S_2	Interrupt
if E **then** S_1 **else** S_2 **fi**	Choice
while E **do** S **od**	Loop
E	Data Expression
Skip	Empty Behavior
delay E	Time Synchronization

The send and receive statements formalize state transitions or events related to the communication of information between components. As an extension to standard UML, message passing in SHE can be conditional. Communication only occurs when the message names match, the number of message parameters is equal and the (optional) *reception condition* E_{rc} (possibly depending on the received data) evaluates to **true**. Both the send and receive statement may be followed by an atomically executed expression E_{at} on data objects.

Utilizing the statement for non-deterministic selection results in executing one of its constituent statements, choosing for the (non-blocking) alternative that performs an execution step first. It is especially useful for formalizing transitions to alternative states in the case where the transitions that can be taken are triggered by different events. Guarded execution allows blocking the execution of a statement S as long as the guarding expression E evaluates to **false** and provides the means for formalizing guard conditions in state diagrams.

Abnormal termination of behavior as specified in state diagrams can be formalized with the abort statement. Next to exceptions resulting in terminating behavior, SHE allows to specify the interruption of behavior. Such exceptions can be formalized by using the interrupt statement. The statements for formalizing a choice or loop have their usual interpretation. We remark that in case the involved methods do not involve return variables, iterating transitions are commonly formalized based on tail-recursive method calls. Data expressions enable formalizing complex functional operations or calculations on data objects. The **skip** statement provides a means for formalizing a transition without any effect.

The statement **delay** E models postponing activity for E units of time. It offers the only way to express quantitative timing behavior. This is sufficient because it can be combined with the interrupt and abort statements to formalize more intricate timing behavior like time-outs or watchdogs. We remark that the time domain of SHE can be dense (in which case E can be real-valued).

SHE provides only a limited number of heuristics for deriving appropriate POOSL code from state or activity diagrams. Using separate methods for representing individual (composite) states or activities is one of the possibilities to obtain such POOSL code. Based on the mathematical approval of combining statements without limitations, POOSL is expressive enough for formalizing behavior according to the designer's preferences.

4. PERFORMANCE EVALUATION

To enable evaluation of performance properties with POOSL, probabilistic behavior and performance related information of the system has to be formalized. For specifying probabilistic behavior like state transition probabilities in state diagrams or probabilistic time delays, SHE provides library (data) classes, which enable instantiating random variables of various probability distributions. Performance related information may explicitly be specified in UML diagrams as QoS characteristics like proposed in [4] or may originate from the involved performance questions. To formalize performance related information, the SHE method involves extending a POOSL model with additional variables and behavior. Such variables enable collecting rewards for the performance metrics that need to be evaluated. To analyze for example the average occupancy of a buffer, we can introduce a variable Occupancy that represents the current occupancy of the buffer. Each time an item is put into or removed from the buffer, added behavior updates Occupancy with a new value. The long-run average occupancy can now be evaluated based on the consecutive values of the variable Occupancy using appropriate mathematical analysis techniques.

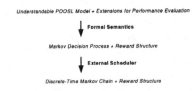

Figure 4. Mathematical framework for performance evaluation with POOSL.

Figure 4 gives a brief overview of the mathematical framework for performance evaluation with POOSL. Based on the formal semantics, each POOSL model defines a unique timed probabilistic labeled transition system [1]. The transition system concerns a Markov decision process, which can be transformed into a *discrete-time Markov chain*. As discussed in [21], the transformation involves resolving any non-determinism in the model by an external scheduler (i.e., execution engine) to ensure that each performance metric gives rise to a single performance figure. The schedulers of the tools presented in section 5 resolve non-determinism in a (uniform) probabilistic manner.

Adding variables to represent rewards for evaluating performance properties results in defining a *reward structure*. This reward structure enables evaluating performance metrics by analyzing the long-run average behavior of the underlying Markov chain. Standard techniques like equilibrium analysis [20] provide the means for *computing* performance results analytically. However, real-life systems commonly entail many concurrent activities resulting in huge Markov chains that are mathematically intractable. Performance evaluation is therefore often based on simulations, which *estimate* the actual performance results. A problem is however that it is unclear how long a simulation should run before the performance results are sufficiently accurate. Estimated performance results therefore only have a proper meaning if their accuracy is known, see for example [10]. Based on the formal semantics, POOSL enables integral accuracy analysis using *confidence intervals*. In [19], we introduced library (data) classes to support the accuracy analysis of different types of long-run averages and variances. These classes additionally enable automatic termination of a simulation when the results become sufficiently accurate.

5.   TOOL SUPPORT

SHE supports an efficient application of POOSL for performance modeling with two different tools. In the formalization stage, the SHESim tool [3] enables incremental construction of POOSL models, that can be validated by interactive simulations. The recently developed Rotalumis tool [1] is intended for high-speed execution of (very large*) POOSL models in the evaluation stage. We briefly discuss these tools below.

In the formulation stage, SHE currently relies on commercial tools for drawing (stereotyped) UML diagrams. Constructing a POOSL model from these UML diagrams is performed by hand. Future research includes an investigation on the possibility of deriving POOSL models more automatically from a consistent set of (stereotyped) UML diagrams specified with a tool that integrates computer support for all three stages of system-level design.

5.1   Formalization with SHESim

The SHESim tool is used for specifying data and process classes with appropriate POOSL code. Cluster classes are defined by drawing instance structure diagrams. A snapshot of the SHESim tool regarding the editing of data, process and cluster classes is given in Figure 5. Notice the resemblance between the instance structure diagram of Figure 3 and the cluster class browser window.

Next to the construction of POOSL models, SHESim supports their execution. Execution of POOSL models boils down to traversing one of the paths through the underlying Markov chain. With the various buttons at the bottom of the system-level editor window in Figure 5, the model can be executed in several modes. As discussed in [3], SHESim allows executing POOSL models on a per scenario basis, which provides a means to validate whether for example different use cases are adequately formalized.

---

* We have been able to analyze models with over a million concurrent activities.

When executing a POOSL model with the SHESim tool, messages passed between processes and clusters are shown on the involved channels. This is illustrated in the lower window of figure 6 and allows validation of the model by comparing the occurring sequences of messages with the specification in collaboration diagrams. In addition, *interaction diagrams* can be generated, which provide ample means to validate the model against sequence diagrams. Figure 6 shows an example interaction diagram.

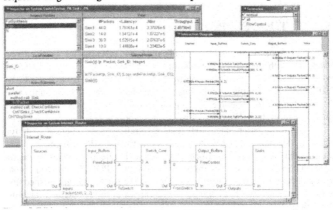

Figure 5. Editing data, process and cluster classes with SHESim.

As indicated in [3], it is possible to open inspectors on each part of an executing model (data objects, processes, clusters and channels). Such inspectors show for example the data object assigned to a variable. As such, they can reflect the (intermediate) results for a performance metric based on using the POOSL library classes of [19], see Figure 6 for an example. It is also possible to inspect the current state of a process. As shown in Figure 6, the statements that can potentially be executed are highlighted. Inspection of processes therefore allows validating whether state and activity diagrams are formalized in an adequate way.

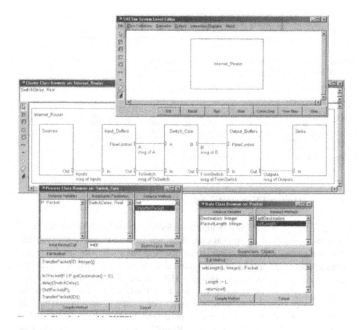

5.2   Evaluation with Rotalumis

Rotalumis is especially developed for high-speed execution of very large POOSL models in the evaluation stage. Whereas SHESim executes a POOSL model in an interpretive way, Rotalumis compiles the POOSL model into intermediate byte code that is executed on a virtual machine implemented in C++. Compared to SHESim, this improves the execution speed by a factor of about 100. Figure 7 depicts a snapshot of the screen output that Rotalumis produces during execution of a POOSL model. It displays only a restricted amount of information about the progress of a simulation. For instance, it shows the run time (RT) of the execution and the simulated or model time (ST). Simulation results for evaluated performance metrics are logged to files. For this purpose, the POOSL library classes for accuracy analysis of performance results introduced in [19] implement logging facilities.

```
*****
****  Rotalumis high-speed execution engine
***   Programmed by L.J. van Bokhoven. (c) 2001
**    For more information visit: http://www.ics.ele.tue.nl/~lvbokhov
*

Garbage collector: incremental Baker's treadmill
Simulation time representation: 64-bit floating point.

Loading & compiling POOSL specification...

Running simulation...

RT: 0.22 Steps: 32769  ST: 0.0011098  GCCC: 0 VMC: 1.12746e+006
RT: 0.45 Steps: 65538  ST: 0.00194425 GCCC: 0 VMC: 2.21533e+006
RT: 0.66 Steps: 98307  ST: 0.00283991 GCCC: 0 VMC: 3.33673e+006
RT: 0.98 Steps: 131076 ST: 0.00377648 GCCC: 0 VMC: 4.4428e+006
RT: 1.37 Steps: 163845 ST: 0.00457058 GCCC: 0 VMC: 5.52231e+006
```

Figure 7. High-speed execution with Rotalumis.

6.   RELATED RESEARCH

In recent years, much research is performed on methods for UML-based performance modeling of real-time distributed hardware/software systems. An important issue is how accompanying tools handle probabilism, concurrency and time. In POOSL, these aspects are inherent to a model and the formal semantics defines how the SHESim and Rotalumis tools manage them [1]. Tools of other UML-based methods often do not separate simulated or model time from run time. For example, deriving SDL [22] specifications from UML diagrams or approaches based on ROOM [17] relying on tools like ObjectTime or RoseRT suffer from the problem that hard real-time behavior cannot be modeled adequately [7]. This is because time is not an inherent aspect of SDL and ROOM models. As a consequence, these approaches are less suited for adequately analyzing for example dedicated hardware systems. Probabilistic behavior can be specified in SDL and ROOM using a proper random number generator. However, ROOM lacks a mathematical basis to enable integral application of performance analysis techniques as is possible when using POOSL, which does incorporate probabilistic features in its formal semantics [1].

The need for a rigorous mathematical basis to support performance modeling was also recognized in [12]. An important conclusion of [12] is the necessity to express probabilistic information in UML diagrams. Both [12] and [15] suggest to develop an integrated framework for explicitly representing performance related information in UML diagrams (as is now standardized in [4]) and the transformation of these UML diagrams into a mathematically analyzable performance model. An example of such an integrated framework is discussed in [5], where the transformation to a queuing network is based on an intermediate textual representation of the performance related information specified in UML diagrams. The SHE method is comparable to [5] in that it also uses an intermediate modeling language to formalize a UML specification. Opposed to POOSL, the language used in [5] lacks however proper support for specifying concurrent activities.

In [12], several options are assessed for automatically deriving a queuing network, Petri net or Markov chain from UML diagrams. References [9, 6] and [11] respectively discuss these options in more detail. In [9], the structure of queuing networks is obtained from use case and deployment diagrams, whereas the behavior of servers in these queuing networks is determined by combining collaboration and state diagrams. However, since the use of exponential distributions in queuing networks only allows for very abstract performance modeling, their practical usability for performance modeling of real-life industrial systems might be limited. In [6], the difficulty of deriving a single Petri net when combining the individual Petri nets of communicating active objects was discussed. The approach in [11] for generating a Markov chain suffered from a similar problem, which was solved using a process algebra. The formal semantics of POOSL combines ideas of traditional imperative object-oriented programming languages with a probabilistic real-time version of the process algebra CCS [8]. Instead of directly deriving a Markov chain as proposed in [11], SHE involves using an understandable modeling language to implicitly define a Markov chain. Only in case of calculating performance results analytically it is really necessary to generate this Markov chain explicitly from the POOSL model. Evaluating performance metrics by simulation merely requires to traverse one of the paths through the Markov chain [19]. This approach is especially efficient for large real-life industrial systems.

7.   CONCLUSIONS AND FUTURE WORK

This paper discussed UML-based performance modeling according to the SHE method for system-level design of complex real-time distributed hardware/software systems. SHE structures performance modeling in the stages of formulation, formalization and evaluation. It uses UML diagrams in the formulation stage for documenting the system's concepts and accompanying requirements. These diagrams comply with an intuitively applicable UML profile, which distinguishes data, process and cluster classes to specify the passive and active resources of hardware/software systems. Basic active resources, of which the behavior is commonly expressed in collaboration, sequence, state or activity diagrams, are modeled with processes. To define the architectural structure of composite active resources, which is often expressed in deployment, component or collaboration diagrams, SHE involves deriving scenario or use case independent instance structure diagrams to define cluster classes.

The UML profile for SHE smoothens utilization of the modeling language POOSL in the formalization stage, where the set of UML diagrams is unified into an executable model. Formalizing the behavior of processes is based on a small set of powerful statements. The formal semantics of POOSL allows one to combine these statements without limitations, providing ample means for formalizing the behavior expressed in collaboration, sequence, state or activity diagrams. One of the statements allows specifying concurrency within processes next to the usual concurrency among processes, which offers full support for modeling concurrent activities.

To enable evaluation of performance properties, SHE entails extending the POOSL model with reward variables based on the performance related information in the UML diagrams or the involved performance questions. In the evaluation stage, the performance properties are evaluated against the executable model to be able to judge whether the system's concepts satisfy the requirements. Proper application of mathematical analysis techniques is supported by the formal semantics of POOSL. Evaluation of performance properties can be performed either by analytic computation or by empiric simulation. In the latter case, designers do not need to be experts of the applied analysis techniques as the underlying Markov chain is implicitly defined by the POOSL model and the accuracy of different types of long-run averages and variances is analyzed based on using library classes.

The SHE method is accompanied with two different tools. SHESim is a tool for constructing POOSL models that can be validated against the UML diagrams by interactive simulations. The Rotalumis tool enables high-speed execution of very large POOSL models in the evaluation stage. The formal semantics of POOSL prescribes how these tools manage aspects like probabilism, concurrency and time, which are inherent to a POOSL model.

Deriving POOSL models from UML diagrams is currently performed by hand. Future research includes an investigation on the possibility of a more automatic generation of POOSL models using a tool that integrates computer support for all three stages of system-level design.

This research is supported by PROGRESS, the Program for Research on Embedded Systems and Software of the Dutch Organization for Scientific Research NWO, the Dutch Ministry of Economic Affairs and the Technology Foundation STW.

REFERENCES

[1] L.J. van Bokhoven. *Constructive Tool Design for Formal Languages: From Semantics to Executing Models.* PhD thesis, Eindhoven University of Technology, Eindhoven (The Netherlands), 2002.

[2] D. Gajski, F. Vahid, S. Narayan, and J. Gong. *Specification and Design of Embedded Systems.* Prentice-Hall, Englewood Cliffs, New Jersey (U.S.A.), 1994.

[3] M.C.W. Geilen, J.P.M. Voeten, P.H.A. van der Putten, L.J. van Bokhoven, and M.P.J. Stevens. Modeling and Specification Using SHE. *Journal of Computer Languages*, 27 (3): pp. 19-38, December 2001.

[4] Object Management Group. *UML Profile for Schedulability, Performance and Time Specification.* OMG Adopted Specification ptc/02-03-02, Object Management Group, March 2002.

[5] P. Kähkipuro. UML-Based Performance Modeling Framework for Component-Based Distributed Systems. In: R. Dumke (Ed.), *Proceeding of the 2nd International Conference on the Unified Modeling Language (UML'99)*, pp. 167-184. Springer (LNCS vol. 2047), 1999.

[6] P.J.B. King and R.J. Pooley. Using UML to Derive Stochastic Petri Net Models. In: N. Davies and J. Bradley (Eds.), *Proceedings of the 15th UK Performance Engineering Workshop (UKPEW'99)* (Bristol, United Kingdom, July 22-23), pp. 45--56. University of Bristol, Bristol (United Kingdom), 1999.

[7] S. Leue. Specifying Real-Time Requirements for SDL Specifications - A Temporal Logic-based Approach. In: P. Dembinski and M. Sredniawa (Eds.), *Protocol Specification, Testing and Verification XV*, pp. 19-34. Chapman and Hall, London (United Kingdom), 1997.

[8] R. Milner. *Communication and Concurrency.* Prentice-Hall, Englewood Cliffs, New Jersey (U.S.A.), 1989.

[9] R. Mirandola and V. Cortelessa. UML Based Performance Modeling of Distributed Systems. In: A. Evans, S. Kent, and B. Selic (Eds.), *Proceedings of the 3rd Conference on the Unified Modeling Language (UML'00)*, pp. 178-193. Springer (LNCS vol. 1939), 2000.

[10] K. Pawlikowski, H.D.J. Jeong, and J.S.R. Lee. On Credibility of Simulation Studies of Telecommunication Networks. *IEEE Communications Magazine*, 40 (1): pp. 132-139, 2002.

[11] R.J. Pooley. Using UML to Derive Stochastic Process Algebra Models. In: N. Davies and J. Bradley (Eds.), *Proceedings of the 15th UK Performance Engineering Workshop (UKPEW'99)* (Bristol, United Kingdom, July 22-23), pp. 23-34. University of Bristol, Bristol (United Kingdom), 1999.

[12] R.J. Pooley and P.J.B. King. The Unified Modeling Language and Performance Engineering. *IEE Proceedings - Software*, 146 (1): pp. 2-10, February 1999.

[13] P.H.A. van der Putten and J.P.M. Voeten. *Specification of Reactive Hardware/Software Systems.* PhD thesis, Eindhoven University of Technology, Eindhoven (The Netherlands), 1997.

[14] J. Rumbaugh, I. Jacobson, and G. Booch. *The Unified Modeling Language Reference Manual.* Addison-Wesley, Amsterdam (The Netherlands), 1999.

[15] Schmietendorf and E. Dimitrov. Possibilities of Performance Modeling with UML. In: R. Dumke (Ed.), *Proceedings of the 4th Conference on the Unified Modeling Language (UML'01)*, pp. 78-95. Springer (LNCS vol. 2047), 2001.

[16] Selic. The Real-Time UML Standard: Definition and Application. In: B. Werner (Ed.), *Proceedings of the 2002 Design, Automation and Test in Europe Conference and Exhibition (DATE'02)*, pp. 770-772. IEEE Computer Society, Los Alamitos (U.S.A.), 2002.

[17] Selic, G. Gullekson, and P. Ward. *Real-Time Object-Oriented Modeling*. Wiley and Sons, New York (U.S.A.), 1994.

[18] B.D. Theelen, J.P.M. Voeten, L.J. van Bokhoven, P.H.A. van der Putten, A.M.M. Niemegeers, and G.G. Jong. Performance Modeling in the Large: A Case Study. In: N. Giambiasi and C. Frydman (Eds.), *Proceedings of the 13th European Simulation Symposium (ESS'01)* (Marseille, France, October 18-21), pp. 174-181. SCS-Europe, Ghent (Belgium), 2001.

[19] B.D. Theelen, J.P.M. Voeten, and Y. Pribadi. Accuracy Analysis of Long-run Average Performance Metrics. In: F. Karelse (Ed.), *Proceedings of PROGRESS'01* (Veldhoven, The Netherlands, October 18), pp. 261-269. STW Technology Foundation, Utrecht (The Netherlands), 2001.

[20] H.C. Tijms. *Stochastic Models; An Algorithmic Approach*. John Wiley & Sons, Chichester (England), 1994.

[21] J.P.M. Voeten. Performance Evaluation with Temporal Rewards. *Journal of Performance Evaluation*, 50 (**2/3**): pp. 189-218, 2002.

[22] ITU-T Recommendation Z.100. *Specification and Description Language (SDL)*, November 1999.

Chapter 13

SYSTEMC CODE GENERATION FROM UML MODELS

Luciano Baresi, Francesco Bruschi, Elisabetta Di Nitto, Donatella Sciuto
Dipartimento di Elettronica e Informazione, Politecnico di Milano, Piazza Leonardo da Vinci 32, 20133 Milano, Italy.

E-mail: *[baresi, bruschi, dinitto, sciuto]@elet.polimi.it*

1.   INTRODUCTION

UML [8] is a specification and design language born from the union of different formalisms in the field of object-oriented design. UML is currently well known and widely used in software engineering. Moreover, its generality has stimulated many designers and analysts that work in different fields to study the possibility of applying UML in their areas of interest, such as, for instance, business planning, system engineering, web design, database modelling.

Some of the main advantages of using UML in an analysis and design flow are:

ease of use, due to the adoption of graphical formalism;

modelling exchange capabilities: UML is highly standardized, together with its extension mechanisms. This allows for model exchanging among different tools.

In the field of the embedded hardware/software system design, the need of tools and languages that allow for an easier management of the ever-growing complexity of such devices is evident.

In particular, the possibility of facing the design task starting at the system level is a point of growing importance.

E. Villar and J. Mermet (eds.), System Specification and Design Languages, 161–171.
© *2003 Kluwer Academic Publishers.*

This claim, in turn, implies some others: the need to model the system at various levels of abstraction, using different formalisms and models of computation; possibility of representing the communication among different model elements at different detail levels; SystemC 2.0 [10], for instance, through its **channel** concept, allows the description of the communication at different levels, thus allowing both the co-simulation of the system in the early phases of the design and the subsequent channel refinement toward the implementation.

The possibility of generating code from UML models (and more generally, from other visual development tools) is very important in smoothing the design process [2]: typically, the design task starts from semi-formal specifications that can be then translated into a set of more formal, precise UML diagrams. The compliance of these diagrams with the original specifications can often be tested directly with the customer, before starting with the implementation. UML does not come along with a specific process, i.e. there is no predefined way of using its diagrams in a design flow. Typically, a set of scenarios describing the system-environment is modelled (or it comes with the specifications); then some hypothesis on the internal structure of the system to be realized are made, by means of functional decomposition or by some architectural constraint (e.g., the system is made of some sensors interacting with a microcontroller, etc.). The scenarios are then refined taking into account the decomposition made. This means that the communication behaviour of the elements constituting the system is investigated and defined. After such modelling steps, the UML model contains information that could be used to generate code consistent with the UML specification obtained.

The expression "code generation" is not unique in the UML world: with this term different strategies are meant [2]. The different code generation approaches can be classified as **partial** and **full**. Again, **full** approaches can be further divided into **generative** and **transformational**.

The **partial** code generation technique starts from a model and aims at obtaining a code skeleton. This means that the code obtained is not executable itself, but defines a framework that captures a wide variety of static and structural aspects. In most cases, the skeleton code generated is a great part, on a quantitative basis, of the final model implementation.

The **full** code generation techniques aim at the generation of an executable model. However, often the UML model does not contain all the information needed to build an executable model (that is, the **behavior** of the model is not uniquely defined). There are two ways to deal with this drawback: one is the **generative** approach, in which the missing details are automatically generated, while the other one is the **transformational** approach, in which the details are manually added to the model.

1.1　Our proposal

An interesting question is whether the modelling capabilities of UML can be applied to embedded design, and integrated in a flow that comprises SystemC 2.0 as the modelling language. In particular, such a flow should allow the use of the high-level modelling features of UML in the early phases of the design, and then it should be able to map this information onto a SystemC model. The advantages coming from such an approach would be of many kinds. Most of them are well proven in the software design field:

- using a visual design approach lets the designer focus on the essential architectural and functional features of the system in the early phases of the project, without being bothered with the many details (syntactical, for instance) of a textual design language;
- visual models are a documentation mean of proven effectiveness; the project documentation task can thus be fairly simplified by the adoption of a visual modelling approach;
- a point of great importance is the possibility, given by a modelling language such as UML, to locate architectural patterns and express them for further reuse; the pattern idea extends the reusability concept from the object to the architecture domain, and is becoming widely used in the design of complex software systems; a great advantage coming from the integration of UML in the design of hw/sw systems would be to explore if such concepts are meaningful in the embedded systems design context; this would be of great interest for the management of the ever growing complexity of the design of this kind of systems.

Among the proposals of UML use in the embedded design area, Rose RT [4][6] defines a profile whose elements' semantics is specified in order to get an executable specification. Instead, our work is focused on giving a set of modelling concepts that can be used to capture high-level aspects of the system being implemented, rather than completely define its executable behaviour. Nevertheless we plan to deal with modelling of behavioural features in the future.

Accordingly, we choose a partial code generation approach. In particular, we want to generate a skeleton code that contains all the structures needed by the system elements to communicate (channels, interfaces, signals, events, and so on).

One of the assumptions of our work is that in the design of embedded systems the code that can be generated from a description of the communication framework is a significant part of the project. With communication framework we mean the information that defines how the parts of the system exchange messages, which in turn is a result of the UML design phase.

What stated above is true in the object-oriented design field, where great relevance is given to the proper definition of the architectural structure of the system (i.e. the modularisation, the interfaces definition and so on). We think that the growing importance that object orientation is gaining in the embedded design field makes this assertion more than reasonable.

The work is divided in two phases:

1- definition of a UML profile, that is, a set of specialized elements useful in the system modelling task. Main characteristics of the profile are:

o expressive power: the elements of the profile have to capture the essential features of the system that is being modelled;

o correlation with the concepts used in other UML diagrams; this should guarantee a smooth transition from the first modelling phases, performed using UML tools such as sequence and collaboration diagrams, to the model refinement using the concept defined in the profile;

o extendibility: the defined concepts should allow further extensions, in order, for instance, to specify behavioral features of the objects being modelled;

2- design of a code generation flow that, starting from a system modelled using the concepts defined by the profile, generates a SystemC code skeleton. The proposed flow is characterized by:

o independence from any specific UML modelling tool;

o flexible architecture, i.e. the code generation is performed in different phases; the result of each phase is be expressed in a format readable by a variety of tools; we choose XML [12] as the intermediate metaformat due to its growing diffusion and to its ease of use.

1.2   Profile definition

The profile has been defined to satisfy the characteristics previously required, by exploiting the **stereotype** extension mechanism of UML. A stereotype is used to tailor UML constructs on the needs of specific application domains.

In Figure 1 the elements of the profile and the relations among them are shown as a UML class diagram.

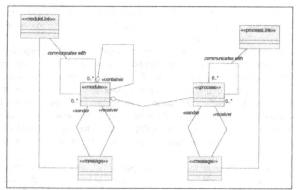

Figure 1 Relations among profile elements

The stereotypes defined correspond to the conceptual entities used to capture the communication features offered by SystemC:

- the <<module>> stereotype is intended, in the profile, as the basic encapsulation element; it essentially acts as a container of processes and of other modules; moreover, the possible communication links among modules are different from those among processes; in this way, the encapsulation features typical of the SystemC 2.0 modules are preserved; the modules can act as sender and receiver of messages, and can communicate with other modules by means of <<moduleLink>> associations.
- The <<process>> stereotype represent the behavioral elements of modules; two processes can communicate directly only if they belong to the same module; communication between two processes of different modules is achieved by means of intermodule communication links; the processes can act as sender and receiver of messages, which in turn are realized by <<processLink>> associations; the <<process>> stereotype is the top of a hierarchy that comprises elements corresponding to the SC_THREAD, SC_THREAD, SC_CTHREAD SystemC process types.
- The <<message>> stereotype is used to represent information exchange between different modules and processes. These are the direct link with the collaboration diagrams obtained from the UML design phases: for every message between two entities in the collaboration or sequence diagrams, there has to be a corresponding <<message>> in the class diagram. Messages must be associated to <<moduleLink>> or <<processLink>> classes, according to the

nature of their senders and receivers (either modules or processes). This association is the link between the UML collaboration diagrams and their SystemC realization.

- <<moduleLink>> stereotype represents the "links" that implement the exchange of a set of messages between two modules. In SystemC this concept corresponds to that of **channel.** In SystemC the channel entity is specialized into less general, lower level specializations: the <<moduleLink>> has the same characteristic. This isomorphism is meant to give control over the code generation phase: the designer can decide to use a signal to realize a set of messages instead of a more general channel; this information will be reflected in the generated code.

- <<processLink>> is analogous to <<moduleLink>>: it represents a communication link between processes belonging to the same module; the main difference is the spectrum of possible implementations of a link: two processes inside a module can communicate with signal sharing, channels, or events that realize simple rendezvous; these possibilities are again represented hierarchically.

2.   PROFILE EXTENDIBILITY

The profile is susceptible of further extensions. These could be the possibility of expressing behavioral features of the model, such as processes finite state machine (FSM) characterization, or communication protocol for channel implementation. A natural way of extending the profile elements is shown in Figure 2.

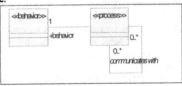

Figure 2 Process behavioral extension

Here the <<process>> class is associated with a <<behavior>> class, which in turn can express some behavioral properties of the process (for instance, it could be associated with a statechart diagram).

The link stereotyped classes (<<processLink>> and <<moduleLink>> classes) are susceptible of a similar extendibility: there could be, for instance, a set of communication protocols that can be attached to a channel and then synthesized in the code generation phase. The profile elements and associations are defined in order to allow such extensions.

2.1 Code generation flow

The proposed design process comprises an UML design phase, a refinement phase that extracts from the UML model the information needed for code generation using the concepts defined in the profile, and two automatic translation phases, that operate a series of transformations to obtain the final code. The implementation of the flow implies the use of different emerging technologies in the field of data exchange:

UML model (collaboration, sequence, class) → UML profiled class diagram: this is the translation phase in which the designer, after having outlined a suitable set of communication scenarios, distils the information needed and expresses it in terms of the concepts defined by the profile. Even though this phase is actually manual for the largest part, a precise methodology can be given and further work will aim at studying the possibility to automate it. Essentially, the steps needed to perform the translation are:

- identify the module-process architecture (i.e., assign every process to a module);
- for each link between two processes in a collaboration diagram:
  - I. define a set of <<processLink>> association classes if the processes belong to the same modules, a set of <<moduleLink>> between the containing modules otherwise;
  - II. assign each <<message>> that connects two processes to a link;
- do the same for each module;

UML profiled class diagram → **XMI model description**: this step is performed by the UML modelling tool; XMI [7] is a XML format that is standardized by OMG [7]; it allows the exchange of design models; XMI provides data exchange not only among UML modelling tools: it has the capability to represent every design model whose metamodel is described in terms of the OMG Meta Object Facility (MOF) [7]. Most of the UML tools now available come along with an XMI generation module, that allows the export of the model in compliance with this XML format;

XMI model description → **XML intermediate format**: the XMI representation of an UML model is very rich of details that relate to things such as the graphical representation of the elements, the references among objects in different diagrams, and so on. Moreover, the data are generated according to the MOF metamodel structure of the UML language: this means that the information associated with the profile elements is not easily accessible. Therefore, this format is not an ideal starting point for the code generation; so, a choice was made to perform a first transformation on the XMI representation, to extract from it only the relevant details needed by the next phases. Another significant choice was to obtain, from this transformation, another XML compliant document. This in fact allows for an easy data parsing by the subsequent algorithms and for much easier data exchange with third parties tools. The technology chosen in order to perform this step is the W3C XLST [11]. XSLT is a set of recommendations for a scripting language that is able to transform a XML document into another XML document by means of a sequence of transformations. The XLST scripts are XML documents themselves. This translation phase is then accomplished by means of an XLST script, whose main tasks are:

- to extract the model information needed to build the intermediate format;
- to format the information retrieved in a useful fashion.

In order to achieve the first goal, the algorithm has to retrieve all the instances of the stereotyped classes, the associations between them, and to output all the related information, formatted as an XML document. This intermediate format contains a list of modules, each one in turn containing a list of processes; for each process there's a set of references to each process that exchanges messages with it, together with the message signatures and the links to what they belong.

XMI intermediate model→SystemC skeleton code: this step can be again performed using XSLT transformations; the intermediate description can also be easily parsed using an XML parser and then elaborated in order to, for instance, compute some metrics from the static information in it contained.

In the sequence diagram of Figure 3 there are three classes that exchange a set of messages. To represent this situation using the concepts defined in the profile the following steps are performed:

1. for each class, decide whether the class is a process or a module;
2. for each process, decide to what module it belongs to;
3. for each couple of processes/module that exchange messages, instantiate one or more links between them (if a process exchanges messages with a module, the link will be between the module and the module that contains the process);
4. assign each message to a link;

A model from the sequence diagram in Figure 3 could be the one represented in Figure 4.

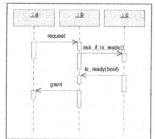

Figure 3 Sequence diagram

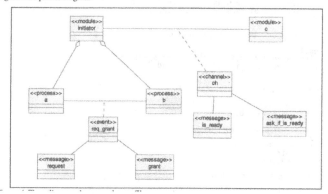

3.   AN APPLICATION EXAMPLE

Let us show, through a simple example, how the proposed flow works and provides as output a SystemC skeleton code starting from a sequence and a class diagram. Even though the benefits of our approach can be appreciable on larger scale scenarios, the example shown is intentionally left straightforward in order to clearly show how the profile items can be used, and the logic that drives the code generation.

Here the following choices have been performed:
–   **a** and **b** are processes that belong to the **initiator** module;

- messages <<request>> and <<grant>> are assigned to the interprocess link req_grant; and <<event>> is chosen in this case to perform the communication;
- c is a module; the messages exchanged with the processes a and b are now exchanged with their containing module;
- messages is_ready and ask_if_is_ready are assigned to the intermodule link ch, which is a generic channel; (in the picture the sender and receiver associations for the messages are not shown for brevity);

```
class ch_initiator_if: public sc_interface
{
    public:
    virtual void send_ask_if_is_ready()= 0;
    virtual bool receive_if_is_ready()=0;
}

class ch_c_if: public sc_interface
{
  public:
  virtual void send_if_is_ready(bool &)=0;
  virtual void receive_ask_if_is_ready()= 0;
}

class initiator: public sc_module
{
    // processes
    void a(); void b();
    //ports
    sc_port<ch_initiator_if> ch_in;
    //events
    sc_event req_grant;
    //constructor
    SC_CTOR(initiator) {...}
};
class c: public sc_module
{
    // ports
    sc_port<c_ch_if> ch_in; [...]
};
```

In the code generated, the modules described in the profiled class diagram have been declared. The interfaces of the modules towards the channel are also generated. If no other information is given, for each message from a module to another a couple of interface functions is defined: one to send the message and one to receive it (the implementation of the semantics of these functions is still left to the designer: they could implement, for instance, a rendezvous, or an asynchronous message exchange). The module **initiator** contains an sc_event, which acts as communication link between the two processes, as specified in the UML diagram in Figure 4. This means that the messages **request** and **grant** are implemented as rendezvous.

4.   CONCLUDING REMARKS

In our work we faced the problem of using UML in the embedded system design with the practical perspective of code generation in mind. We defined a UML profile whose stereotypes can easily capture most of the information contained in the UML sequence and collaboration diagrams; after that, we designed an elaboration flow that generate SystemC 2.0 code skeletons consistent with the UML diagrams. Future effort will concern the definition of profile extensions, together with their effects on the code generation tasks: the models obtained from the addition of behavioural information could be used as a starting point for the exploration of different architectural solutions and, among the same architecture, as a basis for the partitioning task.

5.   REFERENCES

[1] Martin, G.; Lavagno, L.; Louis-Guerin, J. "Embedded UML: a Merger of Real-Time UML and Co-Design", Hardware/Software Codesign, 2001. CODES 2001. Proceedings of the Ninth International Symposium on, 2001 Page(s): 189 -194
[2] Selic, "Executable UML Models and Automatic Code Generation"
[3] Paltor, Lilius "Digital Sound Recorder: A Case Study on Designing Embedded Systems Using the UML Notation", Turku Centre for Computer Science Technical report no 234, January 1999
[4] Lyons, "UML for Real-Time Overview", white paper, Rational
[5] McUmber, W.E.; Cheng, B.H.C. "UML-Based Analysis of Embedded Systems Usign a Mapping to VHDL", High-Assurance Systems Engineering, 1999. Proceedings. 4th IEEE, International Symposium on , 1999 Page(s): 56 -63
[6] Selic, B. and Rumbaugh, J. "Using UML for Modeling Complex Real-Time Systems", white paper, Rational, March 11, 1998.
[7] OMG website: www.omg.com
[8] OMG Unified Modeling Language Specification, version 1.4, OMG specification
[9] OMG XML Metadata Interchange (XMI) Specification, version 1.2, OMG specification
[10] SystemC 2.0 User's Guide
[11] "XSL Transformations (XSLT) Version 1.0", W3C Recommendation 16 November 1999
[12] Extensible Markup Language (XML) 1.0, W3C Recommendation, February 1998

Chapter 14

HARDWARE DEPENDENT SOFTWARE, THE BRIDGE BETWEEN HARDWARE AND SOFTWARE
A Generic Handler Framework

Roel Marichal[1], Gjalt de Jong[2], Piet van der Putten[3], Josephus van Sas[1]
1 Alcatel Bell, 2 Research in Motion, 3 Eindhoven University of Technology

Abstract: Telecom equipment is composed of hardware and software components. Additionally, the test related software on the module, board and system level becomes a bigger relative part of the entire software content. This part gains importance as verification and integration complexity stays increasing at a high rate. This paper describes the characteristics of the embedded SW and more particularly of the hardware dependent software (HdS) design domain in telecom equipment, optimized for reusability, scalability and portability. A basic framework in UML of a generic hardware abstraction layer, from which an HdS platform can be instantiated for any such telecom system, is proposed.

Key words: Hardware dependent software, firmware, telecom, test, shielding, test software, verification, validation, abstraction layer, software

1. INTRODUCTION

In the telecommunication industry, there is a continuous increase in the design complexity. The systems, which are composed of hardware (HW) and software (SW), are getting more complex in a race towards further integration. Additionally, these systems need to be designed with shorter lead-times and life cycles are ever decreasing. Moreover, HW design is becoming more SW focused, giving additional flexibility to be able to cope with changing standards, algorithms and late requirement changes. As a consequence the amount of design effort in SW is increasing relative to the HW design effort. On top of that, the application SW needs to be ported with each new generation of HW, giving also new requirements to the

E. Villar and J. Mermet (eds.), System Specification and Design Languages, 173–182.
© 2003 *Kluwer Academic Publishers.*

architecture of the SW layer just on top of that HW system, which is called HW dependent SW (HdS). The major problem that our design community is facing these days is not only making these complex telecom systems but also verifying and validating these systems. Verification has become the critical issue. Verification of the internals of a programmable System-on-chip is difficult and very labour intensive work. With the co-verification/co-design methodology of programmable System-on-Chips, the HW chip designers start in parallel with the HdS designers who create the bootstrap (bring up of the board/ASIC), the SW drivers for the chip and the test SW for verification and validation of the chip. Through this parallelism the overall product lead-time decreases and the overall quality increases as design bugs are tracked more rapidly. The SW tests must be able to be run in a simulation, co-simulation, co-emulation (SW runs on the workstation, and the HW VHDL is mapped on emulator), emulation and rapid prototyping environment, which are the virtual system environment of the designers. This introduces new requirements towards the lowest HdS layer, as several virtual HW platforms need to be supported and changing from one virtual HW environment to another should be seamless for the SW running on the systems.

This paper presents a basic framework for the SW layer on top of the HW that fulfils these requirements. The outline of this paper is as follows. Chapter two will discuss the different building blocks of which a typical telecom product consists of. The third chapter will explain the life cycle of a telecom system and the different test and SW needs depending on the life cycle phases. Chapter four will explain the need for HdS. The HW abstraction layer, which is a framework that forms a solution to all the requirements that are raised in chapter three will be explained in more detail. Finally in the fifth chapter we conclude.

2.   THE COMPONENTS OF A TELECOM SYSTEM

Telecom systems are composed of HW and SW components. Figure 1 gives a possible static overview on a switching system from a HW *component* point of view. The telecom system is an aggregate of terminals and racks. A rack consists of subracks, cables, lamps and fuses. A subrack consists of functional boards, appliqués and connection boards. A functional board is an aggregate of a bare board and a variety of components (e.g. a chip containing embedded processor cores, etc). In a race towards further integration, the number of components is increasing and the functionality that is put on one ASIC today was put on a board five years ago.

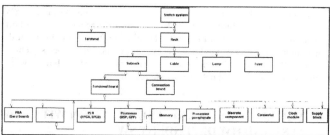

Figure 1. The HW aggregation of a Telecom System (switching system)

Also more programmable devices are present, which leads to a need for a wide variety of tests to verify and validate the integrated system. The *SW components* can be subdivided into three groups: the *application*, the *firmware/embedded* and the *test SW*. The later is a part gaining importance. The test SW is fulfilling a great deal of the verification and validation tasks, by providing complete SW test environment/platform, the embedded SW as well as the functional tests. Bringing up a board/system by taking out of reset, initialising and configuring all the separate components of the system, is done by the embedded SW.

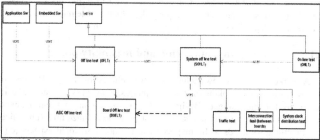

Figure 2. HdS and HdS Test overview

For the tests, we make a distinction between on-line, system off-line and off-line tests. The difference between on-line and off-line is that on-line means doing tests using the operational embedded SW (at ShowTime) with its RTOS, stacks, etc. while off-line means that there are no traffic or other operational things running and used on the device under test that could give interference. The reason not to use operational SW is that often when an error/failure occurs, it is not clear if it is a HW error/failure or a SW error. In order to conclude that it is a HW or SW malfunction, off-line tests are used to indisputably see that the HW fails, without using operational SW. The off-line tests are further subdivided in ASIC off-line tests and board off-line tests. This is shown in Figure 2. The system off-line tests focus on the integration of several modules. They make sure that the system is functionally correct. Several uses associations are also shown, which indicate that tests are as much as possible reused.

3. THE LIFE CYCLE AND THE RELATION WITH THE TEST AND HW DEPENDENT SW

The different major building blocks of a telecom system are identified. Let us now discuss the impact of the different system life cycle phases on the SW architecture and SW test blocks. A use case based analysis is done capturing all the requirements on the architecture of the SW accessing the HW for every phase in the product life cycle. Any telecom system has several phases in its product life cycle starting with i) conception and feasibility, ii) design and verification, iii) Validation and Acceptance, and v) Deployment, field support and phase-out. We will describe all these phases and focus on the impact on the HdS, which is the SW layer just on top of the HW. Instead of the ad-hoc approach often used to write this HdS layer, there is a more structured approach possible, which enable reuse and as such is a source of lead-time and manpower effort reduction.

3.1   The System conception and feasibility phase

A market push or a new technology pull initiates this phase. The results are feasibility studies, plans for the next phases, a business case and a basic go/no go for the project. Design architects search an optimal balance between functional requirements, different possible architectures (on which this functionality needs to be mapped) and the non-functional aspects (e.g. power, area), which are often formulated as constraints. Existing design methodologies and tools are supporting primarily implementation processes for this phase. The requirements for analysis and exploration languages and tools are that they support modeling and give visibility on all functional and architectural aspects. They shall also provide cost functions, which give you an estimate on the non-functional aspects, in order that the design architect can explore and converge towards a balanced optimum. Research is closing the gap [1][2][3] and the industry is using different approaches, working with Rosetta, UML, SystemC, SpecC, etc., to explore the different instantiation possibilities of their system-under-design (SUD). XDSL modems are typically modeled using Matlab and SystemC, to get an idea on the best implementation of the algorithms and the propagation of the quantisation noise. After optimizing, the system architects deliver the executable specifications as well as their system tests. The platform architects then map the application/executable spec on SW and HW architectures. The bridge or borderline between these two is a small SW layer closest to the HW, the HdS.

From a test perspective, it is crucial that the models are checked by system wide tests at this abstraction and that the tests are reused as much as possible or refined if required in the further design phases. By doing this the tests make the design consistent and they capture some key system test knowledge that should not get lost.

3.2   The design and verification phase

In this phase the architecture and its components accompanied by the unit tests coming from the previous phase are further refined. Several methodologies are available (e.g. co-design, ROPES). Depending on the telecom SUD, several components of Figure 1 and 2 will be made. For the design of the HW components there is a close co-operation throughout the design and verification phase between the HW and SW designers, more especially the HdS designers, in the co-verification of an ASIC or FPGA. The HW designers write VHDL/Verilog or generate it by a tool starting from an executable spec [4]. Additionally they write small testbenches to assure basic functionality of the ASIC module. In parallel, HdS designers have set-up their "basic handler" test environment, in which they can do basic actions (e.g. loading binaries, reading registers, running tests). This test environment will actually be part of the delivered product SW and its functionality is reused in all following phases. The HdS designers start writing off-line tests for verification, which run in that basic handler environment.

A key requirement for the SW architecture and the test platform is that the off-line test code does not need to be changed when the (virtual) HW environment, which it is testing, changes. These SW tests must be able to be run in a simulation, co-simulation, co-emulation, emulation and rapid prototyping environment, which are the virtual system environments of the designers. This introduces a requirement towards the lowest HdS layer, as changing from one virtual HW environment to another should be seamless for the test and application SW running on the systems (without porting).

A requirement for the SW coming from a test strategy point of view, is that automatic testing is supported. Every bug fix in HW or SW can introduce new bugs or let undetected bugs surface. The test platform shall support test regression mechanisms.

In case the HW components build for the system are boards circuit boards, the same "basic handler" test environment will still apply. However the designers with whom the HdS designers then interact are the printed circuit board designers. From the moment they are confident that their supply, DC-DC converters, clock circuitry, terminations are all right and they have no shorts, the first thing they most of the time want to see is that they can do some basic actions. For this the basic bootstrap SW of the board (part of the embedded SW) and the "basic handler" general test environment must be provided to him.

3.3   The Validation and Acceptance phase

In the validation phase the designed components must be validated and integrated in the global system. For example, line board integration tests in the telecommunication system as well as global system tests are performed to validate that this is the correct component functionality that had to be made for the customer. A subset of the tests of previous phases is grouped minimizing the coverage overlaps and maximizing reuse. During acceptance the complete telecom system is installed on the site of the client and there the client will do the system tests that he finds necessary to accept the delivery. The same "basic handler" test platform, test cases and the HdS drivers to address the FPGAs, ASICs and some additional made HdS drivers are used during this phase.

3.4   The deployment and field support phase

In the *Deployment phase and Field Support phase the* same "basic handler" test platform, test cases and the HdS drivers to address the FPGAs, ASICs and HdS drivers are used during activities like trouble shooting, logging error conditions and doing all kinds of operation and maintenance activity's. Until the product phases out, the HdS and the tests are used in the field.

4.   HW DEPENDENT SW IN TELECOMMUNICATION

In the previous chapters we identified various opportunities for improvements in the "basic handler" test platform, the test SW and the HW abstraction layer (HAL). The HAL is the access driver layer of the HdS framework just on top of the HW. If the HAL is not reused from the first co-verification phase onwards, it is rewritten and debugged several times by different designers at different times. The ASIC development team is creating basic driver access functions for their testbenches. The SW development teams are trying to make their application SW independent of the underlying HW. The board designers are accessing certain registers they need for testing purposes. The production test SW engineers are creating manufacturing test cases. The application SW engineers are implementing standardized algorithms with test benches and field support people are also reusing the HAL during troubleshooting. Crucial is that all the parties involved state their needs for the HdS HAL in order to maximize reuse. The opportunities of the HAL for reuse and its resulting effort reduction in the HdS layer is clear. We will focus on this layer and propose a basic framework, which shields the SW users of this layer from changes in the HW and vice versa.

4.1   The HW Abstraction Layer framework

The SW components, which are directly dependent on the underlying HW, are the HW drivers, the bootstrap, the download strategy, the built-in tests, the HW dependent parts of the communication provisions, the board support package of the RTOS, etc. The HW abstraction layer (HAL) is the thin SW layer between these SW components and the HW. If the call tree (a tree that shows which function is calling which function) of this HAL SW layer is analyzed you can recognize three levels of functions. On the bottom level you can see two functions, a *read* and a *write* function. They actually do all the accesses towards the HW. The function in the second level on top of this bottom level and who use the read and write function are functions like SetQueueLevel, GetCounter. You could see this as a HW control layer of functions that contain HW access mechanisms towards registers for SW interaction. The highest level of functions, which are using the functions of the second level, form a small application programmer's interface (API) towards the users of the layer. Some examples of this are *init*, which initializes a component, *configure*, which (re)configures a component. Effectively, these functions provide a functional control and communication layer. In the next paragraph the purpose of these three levels of functions will be explained from a shielding point of view.

4.2   The HW Abstraction Layer Shielding hierarchy

The main task of the HAL is to shield HW for upper layer application SW and visa versa. In the shielding there is a certain hierarchy.

The HAL layer is characterised by a framework that is depicted in Figure 3 and it is composed of the following major building blocks: A functional shielding, a register shielding and an access shielding.

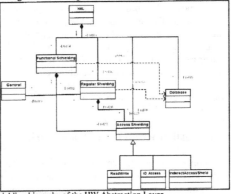

Figure 3. Shielding hierarchy of the HW Abstraction Layer

The *access shielding* has as purpose to shield the actual access to the HW. This is the layer where you map your SW/HdS to the (virtual) HW platform (e.g. emulator in co-emulation). The key goal is that the above SW (e.g. functional shielding, application SW) has no impact of a change of your (virtual) HW platform. The user does not even have to know how the actual physical access is being performed. A list of macros shielding off the actual access of the memory location to allow the use of the HAL in simulations can be used when using C, without loosing performance.

The purpose of the *register shielding* is to make the above SW layers (e.g. functional shielding, application SW) independent of the physical address of a register and other locations that need to be addressed. This shielding enables the above SW layers to only use logical names for these items. For this purpose a *database* is used that does the translation of the logical given names to the actual physical addresses. The database must also contain the particular characteristics on the registers (e.g. ReadWrite (RW), ReadModifyWrite (RMW)). It is also that information that is captured in that database, which is critical for tools that partly automate the HAL layer. In practice for the datatypes structures and/or arrays of basic data types can be used for this.

The *functional shielding* is an HdS layer that actually groups certain functionality in a small, limited API that will be offered to the above SW layers in the response of their needs. The goal of this layer is that by giving a simpler API the abstraction level for the SW designers making higher layers is higher and this eases the comprehension, without requiring in-depth knowledge of the device. This layer is composed of some control code (e.g. state machine), accesses to registers and all other kind of checks. In the next paragraph we will give a basic example of some HAL functions.

```
/* A function of the functional shielding layer*/
Xxx_InitAsicYyy(T_u_int32 * Input, T_u_int32 * Output)
{

T_u_int32  *PF_RegisterX = C_AddressTempPointerLocation;
QE_XXX_HAL_ReadRegisterX(C_ComponentX,PF_RegisterX);

}
/*register shielding*/
T_HdSresult QE_XXX_HAL_ReadRegisterX   (     T_u_int32 ZF_DevNbr,  T_u_int32  *PF_RegisterX  )
{/* Local variables*/
  volatile VCI_T_Module_T_RegisterX *PL_RegisterXBaseAddress;
/*Parameter check*/
  if(ZG_ParamCheck)
  {

    M_HAL_CHECKDEVNBR(ZF_DevNbr, ZG_Component_TotDev);
    M_HAL_CHECKPOINTER(PF_RegisterX);
  }
  /* functionality: Set Addr to RegisterX register */
  PL_RegisterXBaseAddress = &(ZG_pComponent[ZF_DevNbr]-> B_Component.B_RegisterX);
  /*Read RegisterX  and store by client */
/*Field: Value*/
/*access shielding */
  M_RD_REG(PL_RegisterXBaseAddress->Value ,   *PF_RegisterX);
  return E_HDSERR_NO_ERROR;
} /* End */
```

Figure 4. Code example HAL

4.3   A practical HW Abstraction Layer example

Any HW Abstraction layer function is possibly composed of the following code blocks: local variables declarations, parameter checks (to protect against improper use), the functionality itself, clean-up if the functionality manipulated some configurations and then return with a status towards the caller. An example is given in Figure 4 where a function of the functional shielding layer is calling a function from the register-shielding layer. The actual access is shielded through a macro, which takes the correct function call dependent on the (virtual) HW environment you are using at that time in the design phase. You could also use an "inline" approach.

Important is that all accesses from SW towards HW are going through the HAL.

5. CONCLUSIONS

The HdS is a relative new domain with little related work available. Recently the VSI [5] became active with their HdS design-working group. In [6] van der Putten focussed on test development for telecom HW, classifying the HdS test domain by taking different viewpoints. However a generic framework for the HAL was not found in the recent published literature.

The shown HdS approach enables reuse and effort reductions, especially in the HW abstraction layer. The HAL, for which a framework was presented, is the curtain between the HW and the SW. It first is created in the system conception phase and form that moment on starts growing until a full blown driver layer that will stay in the product until phase out. The HAL framework presented offers a means to fulfil all the requirements of a shielding layer of the HW throughout all the life cycle phases of a telecom product. The HAL framework makes your SW design portable and scalable and protects your HW from misuses by the above SW layers. The HdS approach enables reuse and decreases the overall design effort by reusing this HAL as well as all the tests that remain unchanged due to the good shielding of the HAL layer.

ACKNOWLEDGEMENTS

Thanks to the HdS and ASIC designers of Alcatel and the Eindhoven University of Technology's who helped building up these insights.

6.   REFERENCES

[1] F. Catthoor et al., "Code Transformations for Data Transfer and Storage Exploration Preprocessing in Multimedia Processors ", IEEE Design & Test, Vol.18,No 3 No, pp.70-82 , May-June 2001
[2] H. De Man, F. Catthoor, R. Marichal, C. Verdonck, J. Sevenhans, L. Kiss, Filling the Gap Between System Conception and Silicon/Software Implementation, ISSCC2003
[3] Daniel Ragan, Peter Sandborn, Paul Stoaks, A Detailed Cost Model for Concurrent Use With Hardware / Software Co-Design, DAC2002 March 14,2002
[4] http://www.adelantetechnologies.com/
[5] http://www.vsi.org/events/esc02/index.htm : Embedded Systems Conference, San Francisco, 2002
[6] P.H.A. van der Putten, M.P.J. Stevens, G. de Jong, R. Marichal, J van Sas, Test development for telecommunication hardware, ProRISC 2000

PART III: C/C++-BASED SYSTEM DESIGN

Chapter 15

C/C++ BASED SYSTEM DESIGN FLOW USING SPECC, VCC AND SYSTEMC

Lukai Cai; Mike Olivarez; Paul Kritzinger; Daniel Gajski
University of California, Irvine; Motorola; Motorola; University of California, Irvine

Abstract: This paper presents a C/C++ based system design flow that uses SpecC, VCC and SystemC tools. The design starts with a pure C model that is then converted into a SpecC model. A so-called behavior exploration task then takes place to analyze and optimize the system behavior. We then perform architectural exploration using VCC. Once this is complete, the behavior model is refined to an architecture model utilizing the SpecC methodology and the SpecC refinement tool. Finally, the design is linked to implementation using SystemC. We utilize this design flow to achieve the design from C to silicon in an efficient manner. An example of the JPEG encoder is utilized to prove this methodology.

Key words: SpecC, VCC, SystemC

1. INTRODUCTION

There appears to be an increasing trend towards the use of the C/C++ language as a basis for the next generation modeling tools and platform based design methodology to encompass design reuse. However, industry is suffering the pain that there is no one tool and no one complete design methodology that can implement a top-down design methodology from C to silicon.

In this paper we suggest a C/C++ based top-down design flow from C to silicon, to make the system design smooth and efficient, by using currently available tools. We developed our design methodology by using SpecC [1], VCC [2] and SystemC [3]. Although there are some other tools and

185

E. Villar and J. Mermet (eds.), System Specification and Design Languages, 185–194.
© 2003 Kluwer Academic Publishers.

methodologies, such as SPADE, Ptolemy, and Polis, we choose SpecC, VCC and SystemC because they are all C-related and each has strong support in at least one field of design. We made the design flow based on our experiences of attempting to model the JPEG encoder with SpecC, SystemC and VCC, and one internal project, attempting to implement architecture exploration for MPEG encoding and decoding using VCC.

The chapter is organized as follows: section 2 introduces the design flow; section 3 describes the behavior exploration; section 4 describes the architecture exploration; in section 5 model refinements are introduced. Finally, conclusions are given in section 6.

2.   TOP-DOWN DESIGN FLOW

First, we describe the system on three different levels of abstraction:

Behavior level: system model at the behavior level only represents system functionality, without any timing information.

Architecture level: system at the architecture level is described as a set of computation components and storage components connected by system buses. The model of each architecture component is functionally correct with added abstract timing information.

Implementation level: the implementation level describes computation components in terms of register transfers executed in each clock cycle for custom hardware or in terms of instruction sequence for software.

To start design from the behavior level and to end design at the implementation level, designers should accomplish four tasks.

1. *Behavior exploration:* it analyzes and optimizes the behavior model.
2. *Behavior-architecture refinement:* it refines the behavior model to the architecture model.
3. *Architecture exploration:* it allocates system components, maps the behavior to the architecture, and schedule different behavior blocks.
4. *Architecture-communication refinement:* it refines the architecture model to the implementation model.

We use SpecC, VCC, and SystemC together for our design flow, which is shown in Figure 1.

The SpecC methodology is a top down methodology. It provides four well-defined levels of abstraction (models). Among the four models, SpecC *specification model* is at the behavior level while SpecC *communication model* is at the architecture level. SpecC methodology also provides a well-defined method for moving down these successive levels. In terms of assisting in this process, SpecC today provides a profiler at the *specification model*, and a model refinement tool to help in the conversion from *specification model* to *communication model*. We use SpecC for behavior exploration and behavior-architecture refinement.

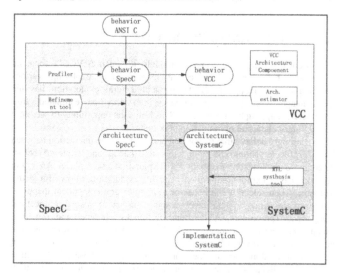

Figure 1. Detailed system design flow

VCC [2] is a behavior/architecture co-design and design export tool. VCC models a system at the behavior level by the use of *whitebox C*, a tool-specific C based language. By mapping the behavior blocks to virtual architecture components saved in the VCC library, VCC estimates the system performance, which helps designers to select the suitable architecture and behavior-architecture mapping solution with the best performance. Therefore, we use VCC to implement architecture exploration.

SystemC [3] is a C++ class library that can be used to create a cycle-accurate model for software algorithms, hardware architectures, and interfaces, related to system-level designs. Compared with SpecC, it has more powerful support for the RTL model of hardware design. Moreover, SystemC co-simulation and synthesis tools can help to generate RTL level design model from the architecture level. Thus, we use SystemC for architecture-implementation refinement.

We easily convert ANSI-C to SpecC and SpecC to SystemC manually, because all of them are C/C++ based. Furthermore, both SpecC and

SystemC support similar system design concepts. To convert SpecC to VCC, we generated a converter to implement the conversion automatically.

3.   BEHAVIOR EXPLORATION

Current platform based design methodologies, such as VCC's, focus on performing behavior to architecture mapping and exploring various combinations of mapping [2]. Purely architectural exploration has the disadvantage that system performance can only be estimated after architecture mapping has taken place. This limits the opportunity to see more global, inter-behavioral optimization opportunities.

We strongly believe that the system behavior and the interaction between behavior blocks should be completely understood and explored before behavior to architecture mapping and exploration takes place. An initial behavior based exploration, called behavior exploration, allows for better heuristics when selecting the behavior to architectural component mapping. For example, since system level design consists of many architectural components (processors, ASICs, and IP blocks), parallel and pipelined execution possibilities among behavior blocks at the behavior level needs to be exploited. However, when using pure platform based design, behavioral exploration is done within the constraints of the chosen architectural components. Because of this, opportunities for performance optimization would be missed by an exploration of different architectural mappings before performance estimation. (In VCC, performance profiling capability is provided by using a built in microprocessor model with associated compiler and profiling tool). These missed opportunities reduce functionality or speed capabilities and optimizations in the final design.

Behavior modeling is not a straightforward task because it must take into account later architecture exploration. Furthermore, behavior modeling for optimized system level design is different from the algorithm modeling associated with pure software design. We also believe a good top-down model for system level design should allow for performance estimation of the design is understood before and after the mapping process takes place.

In this section, we first discuss some confusion between behavior and architecture. We then introduce the behavior exploration.

3.1   Sequential model vs. parallel model

ANSI-C programs consist of a number of functions. The executing sequence among function calls is sequential. Therefore, this is the sequential programming model. On the other hand, hardware language consists of a number of components executing in parallel, which can be called the parallel programming model. Therefore, one step of behavior exploration is the conversion from a sequential programming model to a parallel programming model.

3.2 Behavior parallel, behavior pipeline vs. architecture parallel

Neither the concept of pipelining nor parallelism exists within the C language. However, to efficiently perform behavior modeling, system level design language must supports these concepts. Two terms are defined for this purpose:

a) *Behavior-parallel*: Two behaviors are defined as behavior-parallel if the execution sequence of the two behaviors does not influence the simulation result. Otherwise, the two behaviors are defined as behavior-sequential.

b) *Behavior-pipeline*: If, within a sequential programming model, a number of behaviors are executed one after another in a loop body, and behavior communicates only with the next behavior, then the execution relation between these behaviors can be termed as: behavior-pipeline.

Another term, namely *architecture-parallel* can be defined as: if, during behavior to architecture mapping, behavior A and behavior B are mapped to different architecture components, then the implementation relation between A and B is called architecture-*parallel*. Otherwise it is called *architecture-sequential*.

During behavior-architecture mapping, if we map either a set of behavior-parallel or behavior-pipeline behaviors to different architecture components to form *architecture-parallel*, then we reach a *parallel matching*. *Parallel matching* is a necessary but not a sufficient condition of *parallel execution*.

3.3 Choosing system level design language (SLDL)

First, we need to choose a system level design language for behavior exploration. To understand behaviors by behavior exploration, the chosen SLDL must satisfy three conditions:

a) Ease and accuracy of profiling.

b) Ease of converting from original C language.

c) Ease of converting from one model to another when the execution sequence of behaviors changes.

We evaluated both the SpecC and SystemC languages for behavior exploration by using the JPEG encoder example. SpecC is considered the better language for this task.

3.3.1   Ease and accuracy of profiling

SpecC is a C syntax extension language. SystemC is C++ class library extension. The SpecC behavior model can be profiled relatively easily. It is difficult to accurately profile the SystemC model because of the C++ class library burden, which does not allow for the splitting of system computational needs from the SystemC simulator's computational needs. Furthermore, SpecC provides a behavior profiler.

3.3.2   Ease of model converting

Since SpecC is an extension of the C-language, it inherently supports sequential modeling. In addition, SpecC has the added *par* and *pipe* keywords. This allows for the explicit definition and modeling of behavior-parallel, and behavior-pipeline execution. Converting a C model to a sequential SpecC model is simple because it only contains the syntax change, which is illustrated by the example in *Table 1*. Converting a sequential SpecC model to a parallel SpecC model is also simple because it only contains keyword (*par/pipe*) adding shown in *Table 1*. Similarly, by adding or deleting *par/pipe* keyword, designers can convert one model to another to explore different execution sequences among behaviors.

On the other hand, the SystemC language only supports parallel programming. Since SystemC *processes* are executed in a parallel fashion, it does not support explicit sequential execution. The executing sequences of behaviors are determined by a signal-trigger mechanism. This introduces problems when converting a C model to SystemC and converting from one model to another.

To convert a C model to a sequential SystemC model, designers must keep adding trigger signals and wait statements for each behavior. This is cumbersome and time-consuming, although the top-level behavior of the SystemC shown in *Table 1* does not change much from the C model. For example, to make F1 execute after the execution of F2, designers must declare a signal *F1_done* at the top level behavior, add a statement to trigger *F1_done* at the end of the execution of *F1*, and add a statement to wait *F1-_done* at the beginning of the execution of *F2*.

To convert a sequential SystemC model to a parallel SystemC model, designers must delete old trigger signals and add new signals. For example, parallelization of *F1* and *F2* needs four steps. First, designers delete the signal *F1_done* and related statements inserted in the previous step. Second, designers declare a new signal *F1_F2_start*. Third, designers add a statement to trigger *F1_F2_start* before the execution of the top-level behavior. Finally, designers add statements to wait *F1_F2_start* at the beginning of the execution of *F1* and *F2*, respectively. Since the execution sequence change is implemented by adding/deleting trigger signals and related trigger/wait statements, the processes of converting from one model to another is tedious using SystemC.

Therefore, we choose SpecC rather than SystemC for the behavior exploration.

Table 1. Model example with SpecC and SystemC

Original C model	SpecC model (sequential execution)	SystemC model (sequential execution)
main(){ F1(); F2(); F3(); }	main() { F1.main(); F2.main(); F3.main(); }	sc_main() { F1 F2_inst("inst1"); F2 F2_inst("inst2"); F3 F3_inst("inst3"); }
	SpecC model (parallel execution)	SystemC model (parallel execution)

	main() {
	par {
	  F1.main();
	  F2.main();
	}
	F3.main(); }
		sc_main() {
		F1 F2_inst("inst1");
		F2 F2_inst("inst2");
		F3 F3_inst("inst3"); }

3.4 Behavior exploration process

The tasks of behavior exploration are:

a) Convert a C model to a sequential SpecC model.
b) Determine the granularity of behaviors, merge small behaviors, and split big behaviors based on the profiling results.
c) Explore the specification in order to find *behavior-parallel* and *behavior-pipeline* attributes among behaviors and converting the SpecC model by specifying parallel/pipeline execution relationships among behaviors.
d) Determine the most time-consuming behaviors by analyzing the profiling results.
e) Change the communication models among behaviors in order to enable *parallel execution*.

In task b), we use the SpecC profiler to produce the characteristics of behaviors. The SpecC profiler estimates characteristics of behaviors by computing sum of weighted operations based on a testbench simulation utilizing a designer-provided weight table. By analyzing these profiling results and the structure of the code, the system designer can determine the granularity of the design. Then behaviors are merged and split accordingly. In task c), we find the behaviors containing *behavior-parallel* and *behavior-pipeline* in the specification by using SpecC profiler's traffic analysis function. Designers then update the original sequential SpecC to specify parallel and pipeline execution, by simply adding the par/pipe keywords. The SpecC profiler is again applied to the updated specification model in order to compute the performance improvement given by parallel/pipeline execution. The model is refined accordingly, with the refinement being repeated until an optimal performance is reached.

Finally, we re-model the communication in task e). As we mentioned before, *parallel-match* is the sufficient condition of the *parallel execution*. *Parallel-match* can guarantee the *parallel execution* if and only if the communication among behaviors is modeled correctly.

(a) Sequential model   (b) Pipeline model

Figure 2. Sequential & Pipeline JPEG encoder model

Figure 2(a) shows an error-modeling example for a JPEG encoder. After the *Handle_data* completes its execution, it sends the output(1) to *DCT*. Similarly, *DCT* sends the output(2) to the *Quantization* block, *Quantization* block sends the output(3) to *Huffman*, and *Huffman* sends the output(4) back to *Quantization*, after their execution. Then the output(4) triggers the output(5) produced by *Quantization*, and the output(5) triggers the output(6) produced by *DCT*. As long as *Handle_data* receives output(6), it starts the execution on the next frame of data. Modeling the communication in this way, the four behaviors are executed sequentially, even they are mapped to four different architectural components and have gained *parallel match*.

To make four behaviors run in a pipeline fashion, we remodel the communication among behaviors shown in Figure 2(b). After successor behavior entity receives input from its predecessor behavior entity, it will immediately send an acknowledge back to the predecessor behavior entity, which will trigger the execution of its predecessor. For example, when *DCT* receives input *H1* from *Handle_data*, it immediately sends the acknowledge

H2 to *Handle_data*. As long as *Handle_data* receives *H2*, it starts execution on the next frame of data.

The resulting behavior model has suitable granularity. It can be imported into the VCC architectural exploration tool directly using VCC's graphical input capabilities, or indirectly using the SpecC-VCC converter.

4. ARCHITECTURE EXPLORATION

In this project, we use VCC as our architecture exploration tool. During architecture exploration, we first choose the system architecture. According to the heuristics provided by behavior exploration under SpecC, we then map behaviors to the architecture in order to achieve *parallel match*. We also map time-consuming functions to specific hardware components for the fast execution time. After behavior-architecture mapping, we evaluate the performance of the implementation by using VCC. If the design requirements are not met, then the process of system architecture selection or the behavior-architecture mapping will be repeated.

5. MODEL REFINEMENTS

After we derive the target architecture and behavior-architecture mapping solution, we use the SpecC refinement tool to automatically refine the SpecC behavior model to SpecC architecture model.

After behavior-architecture refinement, the SpecC architecture model is then translated into a SystemC architecture model. This translation allows for the use of SystemC based synthesis tools. This capability allows designers to take advantage of the vast selection of design and verification tools that currently exist to take care of the structural RTL down to mask creation steps of the flow. Though this is true, SystemC currently has and is working on more capabilities that will allow designs done at higher levels of abstraction to be more effective in creating new products more quickly.

SystemC already supports a well-defined HDL model and will support an RTOS and analog model in the future [3]. Using SystemC, implementation modeling and verification can be completed. Furthermore, the SystemC behavior to RTL synthesis tools can be used to generate an RTL model. Clearly defined constructs for synthesis have been developed for both behavioral and structural SystemC. Upon ensuring the model follows these guidelines, synthesis to the final product can be achieved.

Because SystemC is C based, it is possible to easily implement mixed mode simulations between SpecC and SystemC models. This will allow for quickly adding new functionality for product derivatives, and verifying the product behavior before committing to the complete design. The down side remains that the simulation will only run as fast as the lowest abstraction model level will allow. This method should only be used if the original higher abstraction level model is not available and will take too long to create.

6.   CONCLUSION

In this chapter, we provide C/C++ based system design flow based on the use of the SpecC, VCC, and SystemC. The chapter introduced the concepts of behavior exploration under SpecC. These concepts can be utilized to provide a suitable model and heuristics for later architecture exploration under VCC.

Our methodology is a C language-based methodology, from specification to implementation. Conversion from C to SpecC to SystemC is a logical and straightforward process. IP can be modeled at different levels of abstraction allowing for IP reuse, exchange, and integration to be possible. SpecC and SystemC languages enable tools to have a common framework for interoperability. This allows designers to utilize the best "point tool" solution for the implementation of the methodology. Since both languages use C as the underlying technology, interoperability can be achieved. This method quickly converts the C model to an implementation, resulting in decreased design cycle time.

In the design flow, the SpecC profiler, VCC architectural exploration tool, SpecC refinement tool and SystemC synthesis tools, all help to complete an optimized system design. Using this methodology, a JPEG encoder has been created and can be placed into current synthesis tools. Although more automation is needed, the methodology proves design decisions and trade-offs can be easily made at higher levels of abstraction, resulting in an easing of the time to market pressure.

7.   REFERENCES

[1] D. Gajski, J. Zhu, et al., "SpecC: Specification Language and Design Methodology", Kluwer Academic Publishers, 2000
[2] www.cadence.com/products/vcc.html
[3] www.systemc.org

Chapter 16

VERIFICATION OF SYNCHRONIZATION IN SPECC DESCRIPTION WITH THE USE OF DIFFERENCE DECISION DIAGRAMS

Thanyapat Sakunkonchak and Masahiro Fujita
University of Tokyo

Abstract: SpecC language is designated to handle the design of entire system from specification to implementation and of hardware/software co-design. In this paper, we introduce an on-going work, which helps verifying the synchronization of events in SpecC. The original SpecC code containing synchronization semantics is parsed and translated into a Boolean SpecC code. The difference decision diagrams (DDDs) is used to verify for event synchronization on boolean SpecC code. The counter example for tracing back to the original source should be given when the verification result turns out to be unsatisfied. Here we introduce our overall idea and preset some preliminary results.

Key words: SpecC language, event synchronization, difference decision diagrams, boolean program.

1.   INTRODUCTION

Semiconductor technology has been growing rapidly, and entire systems can be realized on single LSIs as embedded systems or System-on-a-Chip (SoC). Designing SoC is a process of the whole system design flow from specification to implementation which is also a process of both hardware and software development. SpecC [1,2] has been proposed as the standard system-level design language based on C programming language, which covers the design levels from specification to behaviors. It can describe both

195

E. Villar and J. Mermet (eds.), System Specification and Design Languages, 195–205.
© 2003 *Kluwer Academic Publishers.*

software and hardware seamlessly and a useful tool for rapid prototyping as well.

This paper introduces an on-going work that tries to develop a technique for the verification of synchronization issues in SpecC language, a system level description language based on C. Recently the semantics of SpecC has been reviewed and clarified [8]. In this paper, we follow those semantics. In SpecC, expressing behaviors within semantic `par` results in parallel execution of those behaviors. For example, `par{a.main();` `b.main();}` in Figure 1 implies that thread a and b are running concurrently (in parallel). Within behaviors, statements are running in the sequential manner just like C programming language. The timing constraint which must be satisfied for the behavior a is $Tas \le T1s < T1e \le T2s < T2e \le Tae$, where notations s and e stand for starting and ending time, respectively. Note that it is not yet determined that any of "st1→st2→st3", "st3→st1→st2", and "st1→st3→st2" is being scheduled. In this case, an ambiguous result, or even worse, an access violation error could occur since st1 and st3 give the assignment value of the same variable x.

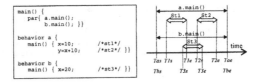

Figure 1. Timing diagram of the threads a and b under the par{ }

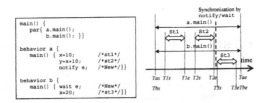

Figure 2. Insertion of synchronization statement notify/wait of Figure 1

The event manipulation statements, such as `notify/wait` could be applied in order to achieve the synchronization of any desired schedulings. `wait` statement suspends the current thread from execution until one of the specified events is notify. The two parallel threads a and b as shown in Figure 2 where the synchronization statements of notify/wait is inserted into Figure 1. The statement wait e in thread b suspends the statement st3 until the specified event e is notified. That is, it is guaranteed that statement st3 is safely executed right after statement st2.

With the timing diagram as shown in the previous figures, we can express the program statements in terms of inequalities of statements timing $Tas \leq T1s < \dots$. Hence, we can make use of the difference decision diagrams (DDDs) [6], a kind of the decision diagram [5], which can represent the inequalities efficiently, in order to verify the synchronization issues of the SpecC programs. SpecC programs are firstly parsed and translated into the boolean SpecC (the boolean programs which are generated from SpecC), then, translate those boolean SpecC into DDD graphs. Original idea of boolean programs were introduced by Ball and Rajamani [7]. The idea here is to abstract any conditions in if statements of the original programs with user-defined *predicates* and translate them into boolean domain. All statements other than event manipulation and conditions for if or switch, and so on, are removed (or abstracted away). Thus only boolean variables and event manipulation statements remain in the generated boolean programs. Here we use boolean programs as a kind of abstracted descriptions from original SpecC descriptions and verify them with DDDs, concentrating on verification of only synchronization issues in SpecC descriptions. Boolean variables are generated based on user-defined *predicates*, which define *abstraction functions* in verification process. Right now we are just assuming that *predicates* are given by designers (who are describing their designs in SpecC), but in the future we plan to develop automatic generation of *predicates* as well.

When verifying the synchronization of SpecC with DDDs, if the result turns out to be true, then verification terminates and the synchronization is satisfied. When the result is false, however, the counter-example must be provided. This counter-example gives the trace back to the unsatisfied source in the original program.

2.   BACKGROUND

In this section, we give an overview of SpecC language, difference decision diagrams, and some basic concepts of the boolean programs. The concepts of sequentiality and concurrency are introduced. Semantics of par which describes the concurrency in SpecC is described as well as the event manipulator notify/wait.

2.1   SpecC Language

The SpecC language has been proposed as a standard system-level design language for adoption in industry and academia. It is promoted for standardization by the SpecC Technology Open Consortium (STOC, http://www.SpecC.org).

Before clarifying the concurrency between statements, we have to define the semantics of sequentiality within a behavior. The definition is as follows. A behavior is defined on a time interval. Sequential statements in a behavior are also defined on time intervals which do not overlap one another and are within the behavior's interval. For example, in Figure 1, the beginning time and ending time of behavior a are Tas and Tae respectively, and those for st1 and st2 are $T1s$, $T1e$, $T2s$ and $T2e$. Then, the only constraint, which must be satisfied, is:

$$Tas <= T1s < T1e <= T2s < T2e <= Tae$$

Statements in a behavior are executed sequentially but not always in continuous ways. That is, a gap may exist between Tas and $T1s$, $T1e$ and $T2s$, and $T2e$ and Tae. The lengths of these gaps are decided in non-deterministic way. Moreover, the lengths of intervals, $(T1e-T1s)$ and $(T2e-T2s)$ are non-deterministic as well.

Concurrency among behaviors are able to handle in SpecC with par{} and notify/wait semantics, see Figure 1 and 2. In a single-running of behaviors, correctness of the result is usually independent of the timing of its execution, and determined solely by the logical correctness of its functions. However, in the parallel-running behaviors, it is often the case that execution timing may have a great affect on the results' correctness. Results can be various depending on how the behaviors are interleaved. Therefore, the synchronization of events is important issue for the system-level design language. The definition of concurrency is as follows. The beginning and ending time of all the behaviors invoked by par statement are the same. Suppose the beginning and ending time of behavior a and b are Tas and Tae, and Tbs and Tbe, respectively. Then, the only constraint, which must be satisfied, is:

$$Tas = Tbs, \; Tae = Tbe$$

According to these sequentiality and concurrency defined in SpecC language, all the constraints in Figure 1 description must be satisfied as follows.

- $Tas <= T1s < T1e <= T2s < T2e <= Tae$ (sequentiality in a)
- $Tbs <= T3s < T3e <= Tbe$ (sequentiality in b)
- $Tas = Tbs, \; Tae = Tbe$ (concurrency between a and b)

The `notify/wait` statements are used for synchronization. `wait` statements suspends their current behavior from execution and keep waiting until one of the specified events is `notify`. Let focus on the `/*New*/` label in Figure 2 of which the event manipulation statements are inserted to that of Figure 1. We can see that `wait e` suspends `st3` until the event e is notified by `notify e`. As for the sequentiality, `notify e` is scheduled right after the completion of `st2` ($T2e<=T_{notify}s$). The only constraint for a single event synchronization is:

$$T_{wait}e<T_{notify}s$$

2.2   Difference Decision Diagrams

The idea of DDDs was introduced by Møller, *et al.* [6]. Its properties are mostly similar to that of BDDs except that it could handle the difference constraints, i.e. inequalities of the form $x-y \leq c$, where x and y are integer or real-valued variables and c is a constant. Figure 3 shows a DDD graph for $\neg(x-z<1) \wedge (x-y \leq 0) \wedge (y-z \leq 2)$. DDDs share many properties with BDDs: 1) they are ordered, 2) they can be reduced making it possible to check for tautology and satisfiability in constant time, and 3) many of the algorithms and techniques for BDDs can be generalized to apply to DDDs. We use these inequalities to represent relating execution timings of event manipulation statements, and use Boolean variables to represent control flows in the SpecC descriptions.

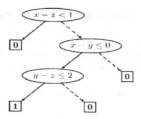

Figure 3. Difference decision diagram

2.3   Boolean Program

The work on Boolean programs has been developing by conduction of Ball and Rajamani at Microsoft Research [7]. They try to conduct the verification on software by realizing the software as a model such that similar to the hardware FSM. This is to make the software model concrete to be verified by using the idea of model checking [3,4]. The Boolean programs have proved to be a subset of the original programs. What distinguishes Boolean programs from FSMs is that the Boolean programs contain procedures with recursion.

As the characteristic that Boolean programs abstract the programs defined by the source language, the satisfied result of verifying some properties on the Boolean programs ensures that those properties are satisfied the original programs as well.

3.   VERIFICATION FLOWS

Our goal is to check whether the given SpecC codes containing concurrent statements par and event manipulation statements notify/wait are properly synchronized. Figure 4 represents the proposed verification flow. We use the idea of the Boolean programs, which represents a subset of the programs defined by the source language, in order to verify for the synchronization of events in SpecC. The flow can be roughly classified into two main stages: verifying and counter-example & refinement stage. Note that the current status of this work is at the verifying stage. We are planning to implement the rest as described in counter-example & refinement stage.

3.1   Verifying Stage (current implementation)

First, the SpecC source code must be parsed and translated into Boolean SpecC code. The Boolean SpecC code contains only conditional (if or switch) and event manipulation statements. Second, the achieving Boolean SpecC is then parsed and translated into the C++ code, which will be incorporate with the DDD package to verify for the event synchronization.

3.1.1   From SpecC to Boolean SpecC

The Boolean programs [7] were proposed for software model checking. It is shown that the model itself is expressive enough to capture the core properties of programs and is amenable to model checking. A similar idea to the Boolean programs is realized to verify for the SpecC synchronization. Let us assume that the original SpecC code to be verified is free of SpecC compilation and syntax errors (let the SpecC language compiler handle this).

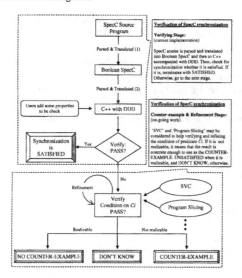

Figure 4. The proposed verification flow

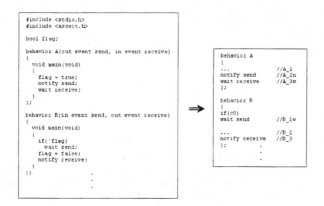

Figure 5. Translation from SpecC code to boolean SpecC

This is to avoid the undesirable results that will occur due to those errors. Then, the SpecC source code is parsing and translating such that

- the event manipulation statements are sustained,

- the conditional statements or predicates of all branching statements are automatically replaced by dummy variables, e.g. `if(x>0)` is replaced by `if(c0)`, `if(x>4)` by `if(c1)`, and so on,

- all other statements are abstracted away by replacing with **skip** (denote in the boolean SpecC by "..." for readability).

```
/***************************************************************/
/* 1) Library header "dddcpp.h" is the DDD library package for C++ */
/***************************************************************/
#include<dddcpp.h>
#include<iostream.h>
#include<stdio.h>
#include<time.h>

/***************************************************************/
/* 2) if/else branching function for DDD expression */
/***************************************************************/
ddd ITE(const ddd& test_expr, const ddd& iftrue, const ddd& iffalse){
  return(test_expr & iftrue) | (test_expr & iffalse);
}

int main(){
  /***************************************************************/
  /* 3) Declare variables with 'boolean' and 'real' type */
  /***************************************************************/
  ddd::Init();
  boolean c0="c0";
  real A_1_a="A_1_a",A_1_b="A_1_b",A_2n_a="A_2n_a",A_2n_b="A_2n_b",A_3w_a="A_3w_a",A_3w_b="A_3w_b";
  real B_1w_a="B_1w_a",B_1w_b="B_1w_b",B_2_a="B_2_a",B_2_b="B_2_b",B_3n_a="B_3n_a",B_3n_b="B_3n_b";

  /***************************************************************/
  /* 4) Generate DDD graphs corresponding to sequentiality and concurrency among behaviors */
  /***************************************************************/
  ddd A = (A_1_a-A_1_b<0)&(A_1_b-A_2n_a<=0)&(A_2n_a-A_2n_b<0)&(A_2n_b-A_3w_a<=0)&(A_3w_a-A_3w_b<0);
  ddd B = ITE(c0,(B_1w_a-B_1w_b<0),False)&(B_1w_b-B_2_a<=0)&(B_2_a-B_2_b<0)&(B_2_b-B_3n_a<=0)&
          (B_3n_a-B_3n_b<0);
  ddd init_AB = (!(A_1_a-B_1w_a<0))&(A_1_a-B_1w_a<=0)&(!(A_3w_b-B_3n_b<0))&(A_3w_b-B_3n_b<=0);
  ddd send = (B_1w_a-A_2n_b<0);
  ddd receive = (A_3w_a-B_3n_b<0);
  ddd init_AB_send_receive = init_AB & A & B & send & receive;

  /***************************************************************/
  /* 5) Verify for synchronization of all events with 'SATISFIABLE' function */
  /***************************************************************/
  if(Satisfiable(init_AB_send_receive))
    cout << "Synchronization is SATISFIED\n";
  else
    cout << "Synchronization is UNSATISFIED\n";

  return(0);
}
```

Figure 6. Translation from Boolean SpecC to C++ with enhanced DDDs

Figure 5 gives an example of translation from SpecC original code to Boolean SpecC code. At the current stage of implementation, we somehow cannot cover all possibilities of parsing and translating the SpecC source, e.g. `for/do/do-while` loops, recursive functions.

3.1.2   From Boolean SpecC to C++ with DDD

When achieving the correct parsed and translated Boolean SpecC code, we again parse and translate to get the outcome in C++ code. Figure 6 gives the C++ code, which translated, from Boolean SpecC in Figure 5. The structure of the generated C++ with enhancement of DDD package is described as followed (from the top down to the bottom of the generated code, respectively).

1. C++ header library where "dddcpp.h" denotes the DDD package library
2. 'ITE()' function which handle the if/else branching
3. Declaration of *boolean* and *real* variables. A_1_a and A_1_b are representing the beginning and ending time of the statement denoted by A_1
4. Constructing DDD graphs
5. Verify for the synchronization using DDD's 'Satisfiable' function

3.1.3   Verifying with DDD

The achieved C++ outcome is then collaborated with DDD package and compiled with C++ compiler to verify for the event synchronization. DDD package provides an ability to check for the satisfiability of the DDD graphs which called 'Satisfiable' function. Users can add properties to be checked. Consider the boolean SpecC in the right box of Figure 5. User may want to check whether the statement labeled B_1w (wait send) is reachable, for example.

The verification result should be 1) true, terminates all processes and returns the synchronization is satisfied or 2) otherwise, continues the process of finding the counter-example.

3.2   Counter-example & Refinement Stage (future plan)

All conditions of if statements in previous stage are abstracted into propositional Boolean variables Ci and we verify for the synchronization without considering about the relationship among those abstracted predicates. Now, we are going to take all those predicates into account to further check whether the unsatisfied result from previous verifying stage can have a kind of counter-example to trace for errors in the source program. In verification at this stage, we are planning to use the following tools to verify and make a refinement of those predicates.

- ***Standford Validity Checker (SVC):*** It is a tool used for validity checking of the boolean formulas. It is proved to be efficient. We will use it for checking for the correctness of the decision procedure.
- ***Program Slicing:*** In general, it is served for finding statements that potentially affect the computation of a specific variable at specific statement. Here we will utilize for some variables tracing.

The conditions on Ci can be considered as:

- "Realizable". This will result in the unsatisfied of the synchronization. The result that we got here cannot be used as the counter-example for error tracing.
- "Not realizable". This means that the result that we have got can really be a counter-example for tracing to the source that give the unsatisfied result.
- "Don't know". The process terminates with "don't know" answer.

4.   CONCLUSION AND OUTLOOK

We proposed the technique for verifying the synchronization of events in SpecC descriptions with the use of DDDs, which is amenable to express the different constraints. The concept of the boolean programs is applied to abstract away some details other than the event manipulation and conditional branching statements. The SpecC code can be checked for the correctness of the event synchronization and let users be able to give some constraints to invoke with the original model from SpecC code. However, up to this point, there are still some limitations on handling the original SpecC code, e.g. the looping, passing variables to a function, recursive functions may not be properly parsed and translated for verification. We are also planning to provide the counter-example and the way to verify it by collaborating with some tools like SVC and Program Slicing.

As a final remark, the proposed technique clearly defines synchronization semantics of SpecC descriptions, which is one of most important issues for system level description languages.

REFERENCES

[1] D. Gajski, J. Zhu, R. Doemer, A. Gerstlauer, and S. Zhao, SpecC: Specification Language and Methodology, *Kluwer Academic Publisher*, March 2000.

[2] A. Gerstlauer, R. Doemer, J. Peng, and D. Gajski,~System Design: A Practical Guide with SpecC, *Kluwer Academic Publisher*, June 2001.

[3] E. M. Clarke, O. Grumberg, and D. Peled, Model Checking, *MIT Press*, January 2000.

[4] K. L. McMillan, Symbolic Model Checking, *Kluwer Academic Publishing*, July 1993.

[5] R. E. Bryant, "Graph-based Algorithms for Boolean Function Manipulation," *IEEE Transactions on Computers.*, Vol. C-35, No. 8, pp. 677-691, August 1986.

[6] J. Møller, J. Lichtenberg, H. R. Anderson, and H. Hulgaard, "Difference Decision Diagrams," *Technical report IT-TR-1999-023, Department of Information Technology, Technical University of Denmark*, February 1999.

[7] T. Ball and S. K. Rajamani, "Boolean Programs: A Model and Process For Software Analysis," Microsoft Research, http://research.microsoft.com/slam

[8] M. Fujita and H. Nakamura, "The Standard SpecC language," *Proc. of ISSS 2001*, Montreal, Canada, October 2001.

Chapter 17

AUTOMATIC GENERATION OF SCHEDULED SYSTEMC MODELS OF EMBEDDED SYSTEMS FROM EXTENDED TASK GRAPHS

Stephan Klaus, Sorin A. Huss and Timo Trautmann
Integrated Circuits and Systems Laboratory, Darmstadt University of Technology

Abstract: Based on an abstract specification denoted as extended Task Graph (eTG) and allocation, binding, and scheduling information, a scheduled and executable SystemC transaction level model (TLM) of distributed embedded systems is generated. Therewith, the whole communication and run-time control is being automatically produced, and the SystemC simulation kernel allows a timely execution and timing validation of the specification. A further benefit of the proposed specification model is the consistent definition of interfaces, which supports the encapsulation of blocks and enables the reuse of IP cores, thus leading to a hierarchical specification of distributed embedded systems.

Key words: Code Generation, SystemC, extended Task Graph, Hierarchical Modeling

1.   INTRODUCTION

More and more complex embedded systems must be built in less time, while also reducing the production costs. Due to this situation, the design process needs to be shifted towards higher levels of abstraction. A distinguishing characteristic of embedded system design is that not only functional behavior, but also non-functional requirements, such as timing, cost, power-consumption, etc., have to be taken into account. In Figure 1 the proposed codesign flow and model code generation is illustrated. After specifying the system with an abstract hierarchical model the optimization steps concerning the fundamental synthesis problems - scheduling, allocation and binding - should be done first. Based on these results an executable hardware and software model is highly desirable, because it allows an assessment of the timed system behavior by means of simulation

E. Villar and J. Mermet (eds.), System Specification and Design Languages, 207–217.
© 2003 *Kluwer Academic Publishers.*

before implementing the core functionality and refining the model. The hardware synthesis process starting from an eTG and resulting in a synthesized gate-level netlist is depicted in Figure 2 for some generic example.

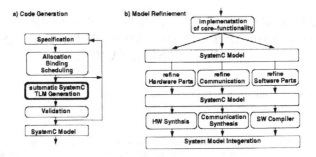

Figure 1. Codesign Flow with Transaction Level Model Generation

As detailed in Figure 1 just one SystemC [10] model framework is necessary to specify the hardware as well as the software parts of the system. When performing gradual refinement of the specification, an executable model is available at any time. Because of the consistent separation of communication and core functionality, this specification approach allows both system optimization and generation of an executable model before implementing any core functionality. The whole communication and run-time control may automatically be generated at transaction level. This yields a validation and behavior assessment instrument and is thus a good starting point for the functional implementation to come. Transaction level modeling covers a higher abstraction level in the modeling hierarchy, whereby hardware-software interactions are defined as processor independent transactions in the first place.

2.   TASK GRAPH MODELS

A specification method for design and validation of embedded systems requires both a suited execution semantic and a consistent notation of both time and concurrency. In [7] accurate definitions of time and different models of computation are emphasized. Task graphs are an appropriate and widely-spread means for the specification of embedded system behavior, because task graphs have a well-defined execution semantic and a temporal order in contrast to, e.g., UML[8] and other highly abstract models. A task graph represents operations and data dependencies between them. In general, task graphs do not address control flow information directly. Therefore, considerable research was done extending task graphs with control flow information as summarized in the following.

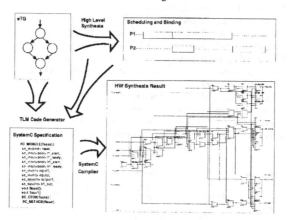

Figure2. Hardware Synthesis using eTG and TLM Code Generator

The CoDesign Model (CDM) introduced in [2] is a structure-oriented approach to data and control flow specification. Input-Output Relations, which introduces behavior classes, are assigned to each process thus capturing the control flow. A similar proposal is the System Property Intervals (SPI) Model [4], where mode tags are introduced to model different behaviors of processes. In addition, input/output data rate intervals and time intervals are associated with each process. In general, the scheduling complexity is the main drawback of both the CDM and the SPI model. They are powerful, but too complex, so that an efficient system analysis in reasonable time is not feasible in many cases. Eles et al. have proposed Conditional Process Graphs (CPG)[3], which model the control flow by means of attributed data flow edges. The value of an associated Boolean condition determines the data dependency, which is actually valid. The extended Task Graph (eTG)[5] enhances CPG by a new class of tasks: A select task is introduced in order to model conditional input behavior without increasing the scheduling complexity. The advocated eTG specification model is summarized in the sequel. Its main features, as outlined above, are both modeling of control flow and a hierarchical structuring of functionality.

Definition (Hierarchical extended Task Graph):

The extended Task Graph (eTG) consists of a directed acyclic graph $G=(V,E)$ and two mappings k,c such that

- $V = I \cup T \cup O$
- E is a set of 2-tuple (v_i, v_j) $v_i, v_j \in V$
- I denotes the set of inputs: $\forall i \in I$. $\neg \exists e \in E$. (v,i) $v \in V$
- O denotes the set of outputs: $\forall o \in O$. $\neg \exists e \in E$. (o,v) $v \in V$
- T are the computational tasks: $\forall t \in T. \exists (t, v_1) \in E \wedge \exists (v_2, t) \in E$. $v_1, v_2 \in V$
- k defines the type of the task: $k: T \to \{normal, fork, join, select\}$
- c specifies conditions: $c: \{e/e \in E \wedge k(source(e))=fork\} \to Conds$.

Conds denotes the set of all possible states in G. An edge $e \in E$ from v_i to v_j implies a dependency between the nodes, e.g., v_j takes data from v_i.

Each task $t \in T$ can be defined recursively as a separate hierarchical eTG denoted $G'=(V',E')$. Each input edge of t $(v,t) \in E$ becomes then an input $i' \in I'$ and each outgoing edge $(t,v) \in E$ becomes an output $o' \in O'$ of G'.

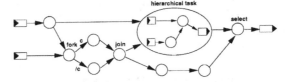

Figure 3 Hierarchical extended Task Graph

An example for the outlined definition of hierarchical eTGs is depicted in Figure 3. A task can start its execution if all required input data is available. Then, after execution, the resulting data are communicated to its data dependent successors. Tasks can be executed in parallel, while the execution of any core program, i.e., the functionality of a task, is of sequential order.

As proposed in [1] any core program needs an encapsulation. So, the core program implements the pure functionality only and both communication and execution order can be managed by one or more control modules. The general architecture of the resulting communicating control units is illustrated in Figure 4. Such a control unit provides an uniform and concise interface to the core programs on one hand and to additional control units on the other hand. Each control unit is also responsible for both the execution order of the tasks it controls and for possibly data buffering. There are two reasons why more than one control unit may become necessary: First, if the specification is partitioned into different pieces of hardware, which embody their own control units, and secondly, if, according to the hierarchical specification, a complex subsystem is implemented on top of IP modules with their own control units.

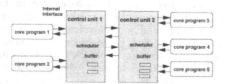

Figure 4 Encapsulation of the Core Programs

3. TLM CODE GENERATOR

The structure of the proposed SystemC code generator and the generator modules are illustrated in Figure 5. Based on the eTG specification of the design object at hand and on allocation, scheduling, and binding selections, three different code sections are generated. First, a code skeleton for each task including data and control signals is coded in SystemC. Secondly, the control units for each hardware module are derived. Thirdly, the top-level entity is created, thus instantiating and connecting all control units and their associated tasks.

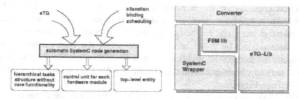

Figure 5. SystemC TLM Code Generator

The code generator supports hardware synthesis due to the fact that the control units are coded in RTL style. A gate-level netlist is produced from the RTL code using the CoCentric SystemC Compiler. This generated netlist can be fed into the traditional hardware synthesis flow. First application examples have been successfully tested be means of this design flow.

3.1   Communication

The complete internal communication is managed by on or more control units as follows. Every task has a port for each data edge and start and ready ports. All these ports are connected automatically to the control unit via signals. At transaction level a communication is represented by a communication time (i.e., defined by the amount of data to be transferred) and a communication medium. So, communication can be handled similar to any other computational task, requiring some execution time and an allocated resource. Therefore, the proposed SystemC code generator produces communication processes. This fits nicely with the SystemC 2.0 communication paradigm. Communication in SystemC is modeled by ports, interfaces and channels. An interface provides a set of method definitions, which constitute the access to the communication medium. Ports are objects through which a module can access an interface, whereas a channel implements the communication functionality. The generator defines the interface to the communication medium using a handshake protocol, but not the communication protocol itself.

3.2   Finite State Machine

The control unit is implemented as a Finite State Machine (FSM). The scheduling and the binding relations are taken to create the FSM, which is subsequently   generated as a SystemC **module**. Each state of the FSM activates one or more core programs, as outlined in Figure 4. Its next-state function depends on the termination of tasks and possible control flow decisions. A **ready**, a **start**, and a **state** signal, respectively, are generated for each task managed by the control unit. In order to implement the control flow as modeled by the conditions in the eTG, the **ready** signal can be enhanced by control information derived from the associated **state** signal. This leads to conditional branches in the FSM state transition diagram as visible from Figure 6.

The FSM representation is generated in a straightforward manner from the execution order of the tasks, which is derived from the scheduling sequence as detailed in Figure 6. The external **start** signal activates the FSM. Thus *T1* and *T2* are started. Upon termination of *T3*, the control flow is split: If the **state** signal from task *T3* is /C then *T4* is activated, otherwise *T5* is activated.

According to Figure 4 embedded systems may embody more than one control unit. As a consequence, these units have to share control information. However, as pointed out in [9], the overall latency of the system may even be improved in case that control information is being shared between different control units.

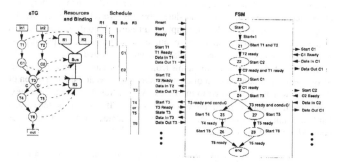

Figure 6. eTG Specification, Binding, Scheduling and resulting Control Unit

3.3 SystemC Code Generation

In this section the approach to automatic SystemC code generation is outlined. Table 1 summarizes a generic task module. Its interface defines **start** and **ready** ports as well as data input and output ports for each task. Tasks are activated by their corresponding start signals.

Table 1. Tasks Module (Left: Interface; Right: SC_THREAD of a Task)

```
SC_MODULE(Tasks) {                          void Tasks::Task3() {
  Sc_in<bool> reset;                          while (true) {
  sc_inout<bool> t1_start;                      while (t3_start.read()==0) wait();
  sc_inout<bool> t1_ready;
  sc_in<int> input1;                            int a= dataT2_T3.read();
  sc_in<int> input2;                            int b= dataT1_T3.read();
  sc_out<int> output;
  sc_out<int> t1_out1;                          // Insert code here:
  void Reset();                                 a=a+b;
  void Task1();
  void HW1();                                   wait(100-1, SC_NS);
  SC_CTOR(Tasks) {
    SC_METHOD(Reset);                           dataT3_T4.write(a);
    sensitive_pos << reset;                     t3_ready.write(1);
    SC_THREAD(Task1);                         };
    sensitive << t1_start;                    }
}
```

The task process waits on its start signal. Then, after activation the process reads data from the input ports. In the generation flow, the core functionality of the processes is initially omitted, whereas only the associated timing behavior is modeled by the **wait**() statement. In the example of Task 3 in Table 1 a simple addition is the only function coded into the process. Then, upon the computation completes, output data is assigned to the output port and the ready signal is activated.

The FSM for the control unit stems from the Huffman model and consists of two parts: The computation of the next state function and the activation of the tasks, which are assigned to the states. As illustrated in Table 2 the next state function may depend on multiple signals. If task $t1$ is ready and hardware unit $h1$ is executed at least once, then state $z3$ can be assigned as detailed in the left part of Table 2.

Table 2. Control Unit (Left: Calculation of Next State Function; Right: Activation of a Task)

```
Switch(actualState) {
  Case ready : {                              Switch(nextState) {
    if (start.read()==1) {                      case z3 : {
      nextState=z1;                               // read output to buffer
      break;                                      buffer1.write(t2_out.read());
    }
  }                                               // write input data from t1
  case z2 : {                                     h1_in.write(t1_out1.read());
    if (t1_ready.read()==1
        && h1_count.read()>=1) {                  // start h1
          nextState=z3;                           h1_start.write(1);
    }                                             break;
    break;                                      }
  }                                           }
}
```

The activation of $h1$ is outlined in the right part of Table 2. First, the output data of $t2$ is moved to a buffer and then the output data of $t1$ is assigned to the input of the next task $h1$. Finally, $h1$ is activated.

4.   RESULTS

An application of the proposed methodology is demonstrated in the sequel by means of a small example. However, the methodology was successfully applied to complex specifications consisting of more than 200 tasks [11]. Besides scheduling, the complexity is linear in the number of tasks and allows handling such complex systems very easily.

Figure 6 depicts the chosen example. The eTG specification consists of 6 tasks, whereas the dashed lines denote the binding for three computational resources and one bus, respectively. The schedule is derived from executing a genetic scheduling algorithm [6]. On the right side in Figure 6 the automatically created state transition graph using eTG, allocation, binding, and scheduling is depicted, which forms the basis of the subsequent code generation. The result of the proposed generation step is a scheduled, executable specification model coded in SystemC. For analysis purposes the timing waveforms of all relevant control and data signals may be produced. Response times as well as signal delays and signal values can be observed and validated. The waveforms of this example are depicted in Figure 7.

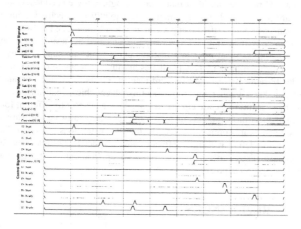

Figure 7 Waveforms of the Control, Data and External Signals

The output data is set at *t=782* to the value of *8* (i.e., the sum of the input values *3* and *5*, respectively). Not just delays of the system, but all other relevant timing information can be derived from such simulation result.

5.   CONCLUSION AND FUTURE WORK

The proposed approach enables system designers to shift the design process to a higher level of abstraction. Our methodology is based on an automatic model code generation based on hierarchical task graphs as an entry point. It is aimed to handle the increasing complexity of embedded systems and thus shortens 'time to market'. Especially the separation of communication and functionality as well as the well-defined activation semantic allow the introduction of uniform control units. As it has been demonstrated by means of an example, the application of the TLM generator to the domain of embedded systems design is highly appropriate. The automatically generated communication and control structures leads to earlier analysis results and error-free code. Having an executable scheduled system model at hand a faster validation of the non-functional requirements becomes possible. Later on developers of embedded systems can concentrate on implementing the functionality of the computational tasks.

SystemC provides an ideal platform for engineering embedded systems. Software as well as hardware modules can jointly be specified using the same language and verified using a common test-bench. The hardware parts may then be refined up to RT level and implemented by means of synthesis tools. The hierarchical modeling features of SystemC are supported by the our specification model. This facilitates not just a structured design, it also enables IP reuse in a straight-forward manner. FSMs can also be organized in a hierarchical manner, thus implementing a hierarchical control flow.

In the near future we are aiming to extend the generator to different forms of communication paradigms including the necessary software-software and hardware-software communication. The intended extensions of SystemC 3.X related to modeling real-time software with abstract RTOS and scheduler services promises further possible fields of applications of the proposed model generation approach.

REFERENCES

[1] W. Boßung, T. Geyer, S.A. Huss, and L. Wehmeyer. Specification and Validation of Information Processing Systems by Process Encapsulation and Symbolic Execution. IEEE CS Workshop on VLSI, Orlando, Florida, April, 2000.

[2] W. Boßung, S.A. Huss, and S. Klaus. High-Level Embedded System Specifications Based on Process Activation Conditions. Journal of VLSI Signal Processing, Special Issue on System Design, Kluwer Academic Publishers, vol. 21, no. 3, pp. 277-291, 1999.

[3] P. Eles, K. Kuchcinski, Z. Peng, A. Doboli, and P. Pop. Scheduling of Conditional Process Graphs for the Synthesis of Embedded Systems. IEEE/ACM Proc. Design, Automation and Test in Europe, pp. 132-138, 1998.

[4] R. Ernst, D. Ziegenbein, K. Richter, L. Thiele, and J. Teich. Hardware/Software Codesign of Embedded Systems - The SPI Workbench. Proc. IEEE Workshop on VLSI, Orlando, USA, June 1999.

[5] S. Klaus and S.A. Huss. Interrelation of specification method and scheduling results in embed-ded system design. In *Proc. ECSI Int. Forum on Design Languages*, Lyon, France, September 2001.

[6] R. Laue. Scheduling mit genetischen Algorithmen. Project Alfa Core(in German) Integrated Circuits and Systems Laboratory, Darmstadt, University of Technology, 2002.

[7] E.A. Lee and A. Sangiovanni-Vincentelli. A Framework for Comparing Models of Computation. IEEE Transactions on CAD, 1998.

[8] OMG. Unified modeling language. Specification v1.3, Object Management Group, Juni 1999. http://www.omg.org/technology/uml/.

[9] P. Pop, P. Eles, Z. Peng, and A. Doboli. Scheduling with bus access optimization for distributed embedded systems. In *IEEE Transactions on VLSI Systems*, volume 8, pages 472–491, October 2000.

[10] SystemC. Functional Specification For SystemC 2.0. http://www.systemc.org, 2001.

[11] Trautmann. T. Entwurf und Implementierung eines SystemC Code Generators basierend auf Task Graphen. Diploma Thesis (in German) Integrated Circuits and Systems Laboratory, Darmstadt, University of Technology, 2002

Chapter 18

SYSTEMC-PLUS COMPLEX DATA TYPES FOR TELECOM APPLICATIONS

Massimo Bombana*, William Fornaciari**, Luigi Pomante***

*Siemens Information and Communication Networks SpA, R&D Development Technologies PT-CA2, Via Monfalcone 1, I-20092 Cinisello Balsamo (MI), Italy. (Massimo.Bombana@icn.siemens.it)

**Dipartimento di Elettronica e Informazione, Politecnico di Milano, Piazza Leonardo da Vinci 32, 20133 Milano, Italy.(fornacia@elet.polimi.it)

***CEFRIEL, Via R. Fucini 2, 20133 Milano, Italy. (pomante@cefriel.it)

1.   INTRODUCTION

The telecom market segment is characterized by a constant search for innovative solutions and by a strong competitive profile. These elements stimulate a continuous identification and formalization of enhanced industrial design flows and methodologies, exploiting design reuse [2,3], extended verification [1], abstraction [4] and power aware techniques. All these efforts share the common goals of improving design practice by:

increasing the abstraction level of HW models using HDLs;

applying early multiple-constraint driven design space exploration;

improving the automatic flow in order to minimize design time and effort.

SystemC [6, 7] and C++ are nowadays considered system design languages that allow design modeling for complex HW modules or mixed systems [5]. A value added to such design modeling strategy is offered by SystemC-Plus [8-10], that is a SystemC extension combining the benefit of

E. Villar and J. Mermet (eds.), System Specification and Design Languages, 219–229.
© 2003 Kluwer Academic Publishers.

adopting an object methodology for system specification and analysis with the possibility to easily fit FPGA-based synthesis flows.

It is clear that one of the most beneficial features of this approach is the possibility to mix various levels of model abstractions as well enhancing reuse of modules during the system-level design phase. The designer can model an item of the system through a pure functional untimed specification, and then check its interaction with other elements of the system in an early design phase.

In order to adopt this framework in industrial sites, two essential key features must be made available: the definition of a reliable design flow and the development of application specific libraries to increase re-use and decrease design time. These elements solve many methodological aspects, flowing into the preparation of design guidelines able to do provide reliable results.

In this paper, we focus on the approach used to cope with such key elements, i.e. the identification of a suitable design flow and the definition of a base library for telecom data types for multi-purpose applications. The paper is organized as follows. In section 2 we summarize the technical features of several HW design flows, based on VHDL and SystemC. In the proposed flow, the specifications in SystemC are successfully coupled to commercial EDA tools down to gate level netlist. In section 3, we describe the modeling strategy adopted to develop the library cells. In section 4, we describe some of the developed data types, and in section 5 how these can be used to model telecom systems. Finally, we provide some conclusions and forecast for future work.

2.   DESIGN METHODOLOGIES

Our main concern is to identify, formalize and apply a complete and automatic ASIC/FPGA design flow suited for telecom applications and with the characteristic of combining design abstraction, behavioral synthesis and extended simulation.

Now, in industrial environments, assessed design practices are still based on VHDL at RT-level. The benefits of adopting behavioral design and synthesis has been recognized but the corresponding EDA tools probably are not totally fulfilling the industrial necessity and/or expectations.

Figure 1 shows three alternative design flows. The first on the left of the figure shows the *state-of-the art* where the system specs are provided (or translated) in VHDL at RT level and through commercial synthesis tools finally synthesized. The output can be a gate level netlist or a FPGA description. The second (center of Figure 1) pulls up the abstraction level by considering as input a VHDL behavioral description and a related synthesis environment.

The latter, on the right side of the figure, is very attractive since SystemC mixes the power of object-oriented analysis with the possibility to deal with HW/SW platforms in an integrated manner. Moreover, industrial community is spurring EDA vendors to provide complete SystemC-based toolsets driving the designers from the system-level abstraction level down to the implementation (e.g. CoCentric [12]).

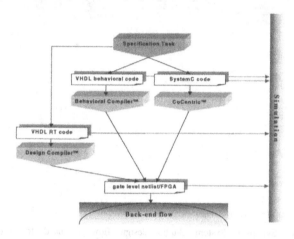

Figure 1. Actual design and tool flow.

The key requirements for the success of this last flow are:

the availability of application-specific library;

verification tools covering all the refinement steps (and abstraction levels);

clear methodological guidelines, especially for the top level design activities.

This paper considers the system flow that is going to be supported within the *ODETTE IST Project*, based on *SystemC-Plus*. It is envisaged the presence of *System Studio* [12] as the behavioral synthesis engine and of a pre-processor allowing the designer to model and refine the application by following a object oriented approach.

We consider SystemC-Plus: a specification language that allows more abstraction and reuse in the description of hardware. Moreover, the language is tailored for synthesis. The design flow in which our designers intend to use it is a *Synopsys-based* synthesis flow, where more expressive specifications can be adopted.

Therefore, our flow includes simulation, synthesis and co-simulation (when needed). In comparison with the flows described in the previous section (see Figure 1), this one has a new layer that sits above the behavioral one (seeFigure 2).

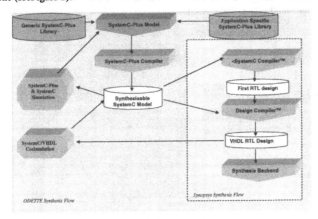

Figure 2. ODETTE Synthesis Flow.

The Synopsys System Studio design flow is used to reach implementation level (right of Figure 2). The input to this flow is provided by *SystemC-Plus Compiler*. This input can be at two different abstraction levels: RT (and in this case it is given to *Design Compiler*) or behavioral (and in this case it is given to *SystemC Compiler*). The abstraction level depends mostly on the adopted style for writing the SystemC-Plus model. The two ODETTE libraries are used as design support in the modeling phase, providing simple pre-defined objects, like data containers, objects for computation, or simple telecom structures like ATM cells or TCP/IP frames.

Two simulation levels are included in the flow (left of Figure 2). The first simulation is at SystemC-Plus and SystemC level, and it verifies the functional description provided by the designers and the version created by SystemC-Plus Compiler. This step also covers verification of the tool functioning (i.e. verification is done prior and after compiling the model into regular SystemC). The second simulation is performed at lower level of the design flow and, after the synthesis, to verify the timing constraints meeting and other similar implementation related details. This phase can include a pure VHDL simulation, or a SystemC/VHDL co-simulation according to the designers needs. In both cases, if the implementation does not comply with the specification, the designers go back one step, respectively to the SystemC-Plus model or to the SystemC model to modify it accordingly before a next synthesis step is performed. Finally, when the implementation meets the specification, the design process reaches the lower design levels, like FPGA programming or ASIC netlist generation. Such a flow can be easily targeted to accommodate current design flow requirements for Telecom applications. Moreover, ASIC/FPGA flow is enough general-purpose to be applied in every market segment where complexity of devices, power consumption and strict time to market are the relevant issues.

3. MODELING OF A SYSTEMC-PLUS LIBRARY

SystemC-Plus extensions of C++ enhance SystemC modeling with three basic features allowing a more abstract description of systems:

specialization of classes through inheritance;

use of polymorphic objects;

use of private interfacing objects and communication.

Each feature shows clear benefits in term of modeling style, mainly based on extended re-use. However, in the telecom domain, the requirements for the manufacturers are focused on synthesis. This imposes a constraint on the library models deriving from the compliance to the subset supported by current (or future) synthesis tools. For instance, limitations in expressiveness due to synthesis are that class templates must contain only parameters of fixed types, the functions can't return reference value, destructors can't be redefined, etc.

Application-specific library elements covering different types of telecom applications (i.e. both for mobile and fixed networks) are more complex than the library cells defined in a generic support library. Therefore, they are less fit to extended reuse.

In an object oriented environment the definition of library cell differs notably from the usual concept of library cell, as applied for instance in a VHDL environment. In fact, a VHDL library element is a model that can be instantiated as it is in the design, providing some parameters that are defined in the GENERIC part of the entity, allowing choosing a few characteristics of the object to be instantiated. In SystemC-Plus two cells can be modeled using specialization or templates. In the former case more specific elements are derived from general ones while using the template feature, different parameters can be associated to a specific implementation.

The latter is very powerful, for instance data containers (FIFOs, Stacks, and so on) that can contain different types of data (bytes, packets, ATM cells, etc.) are modeled using template parameters specifying the type of data of the elements to be manipulated. From our analysis, it was verified that the most expressive strategy should gather both approaches through the definition of class hierarchies (i.e. specialization) using templates. This mixed approach is supported by the SystemC-Plus synthesis toolset.

4.　　TELECOM-ORIENTED ABSTRACT DATA TYPES

Data aggregates like cells, packets, and so on are widely used in many domains of telecom applications. These aggregates are used to transfer structured information on different types of communication channels. Examples of these data aggregates are the records sent on air according to the GSM standard, or the Asynchronous Transfer Mode (ATM) cells. Similar types of packets, called datagrams, are used for Transmission Control Protocol/Internet Protocol (TCP/IP) transfer modes on the Internet network. All these cells are characterized by a fixed or variable structure made up of fields, each one carrying a part of the data information to be transferred. All these data aggregates can be modeled as abstract data types, and they share a great deal of common features and methods, assuring a wide amount of abstraction and reuse, encapsulating knowledge and information encountered with high frequency in these applications.

In the following, the *ADT* related to some components (i.e. ATM cells and IP Datagrams) are described with more details in order to provide significant examples about the contents of the *Application-Specific Library*.

4.1　　ATM Cells

The ATM domain is focusing on elaboration and manipulation of ATM cells, including functions to translate from custom/standard representations (for example company proprietary data format and so on). Moreover, most of the HW devices include complex control sections.

A custom ATM cell is composed of a variable number of fields, dedicated to contain data related to the transferred information and some control parameters. A basic distinction is done between header and information fields. Figure 3 shows as an example the possible structure of a custom (i.e. external plus custom header) ATM cell.

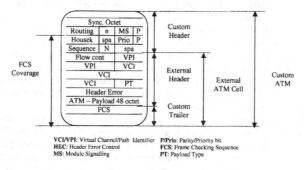

VCI/VPI: Virtual Channel/Path Identifier	**P/Prio**: Parity/Priority bit
HEC: Header Error Control	**FCS**: Frame Checking Sequence
MS: Module Signalling	**PT**: Payload Type

Figure 3. Example of ATM cell structure.

It is natural in this domain to think of modeling ATM cells as abstract data types where methods implement most of the meaningful transformational procedures related to the cell manipulation. Moreover, the *ATM ADT* has been built on the base of other generic classes belonging to the *ODETTE Basic Library* exploiting the reuse and specialization capability provided by the proposed flow. A specialization hierarchy models the ATM cell and template choices enhance the scope of the application domain.

The ATM cell is modeled as an aggregation of three different parts: a header, a payload (the 48 fields containing the information) and, optionally a footer. All these classes are derived from Memory. The *GenericHeader* is an abstract class and it is specialized to describe the standard header of a generic cell. Subclasses of this class increase specialization, i.e. they represent headers with more features than the standard ones. For instance, some attributes of the *CustomHeader* class represent a complex, proprietary structure that may change from Company to Company, including for instance the *Routing*, *Housekeeping* and *Sequencing* fields. Finally, the *Trailer* is also modeled in the same way.

The methods belonging to these classes mainly have the task to set and to read the custom values of the corresponding fields. Methods to perform different types of checks, such as *CRC* (*Cyclic Redundancy Check*) verification, parity generation and test are also implemented. The next section describes with more detail the structure of the *ATM Generic Header*.

4.2   TCP/IP Data Types

The *ADT* defined in the previous sections can be easily extended to other application domains. As an example, we briefly describe here the data types needed for the *Transmission Control Protocol/Internet Protocol (TCP/IP)*.

The Internet reference model, called *4-layer Internet reference model,* is organized into four conceptual layers that build on a fifth layer of hardware. The data structure remains the same throughout the three levels. The application data is the data field in the *TCP* segment. The *TCP* frame is the data field in the *IP Datagram* and so on. The data is simply encapsulated into its own frame structure as the payload.

Unfortunately, the inclusion cannot be interpreted as an aggregation in terms of objects. In fact, different payload sizes are allowed by the standard. That means that it is possible for a single frame at IP level, to be fragmented into several pieces to be sent through the link layer. So no real aggregation can be applied here, but methods that operate on the frames in order to segment them for transmission and to assemble them again. However, for *TCP segments* and *IP packets* structures the analogy with the *ATM cell* in terms of datatype modeling is clear: the basic structure is similar (several fields organized in larger records), even if the meaning of each field is application specific.

5.   USE OF ADT IN TELECOM APPLICATIONS

The main advantage of a library of classes like the ones described in the previous section is to allow the specification of applications where the data path part is decoupled from the control flow. In very complex applications like the ATM or TCP/IP based ones, this is the first step in the task of partitioning, i.e. identifying the sections that can be developed in SW, from those that can be convenient to develop in HW, or apt for specific DSPs, or specific processors like *Network Processors*. This phase is part of the architectural study where alternative solutions can be evaluated and tested to cope complexity.

The simplest paradigm for the description of the control part and for modeling abstract protocol layers, is the hierarchical *Finite State Machines (FSM)*, with states identifying the main functions to be applied on the data types. The state machine has several states. In each of the states, cell manipulation is performed, in order to isolate HEC, do computations and comparisons, information storing and retrieving. If an error is found in several consecutive cells, the machine assumes that synchronization has been lost and returns to the search process.

More complex data manipulations include scrambling and unscrambling.
SystemC-Plus specification task covers different phases:
1.   Specification coding. Using the library objects, a standard SystemC
      module is created. A main program is defined, where the ports of the
      module are declared, and the constructor is defined using
      SC_CTHREAD. The ports at least include I/O channels, a reset port of
      type bool, the clock plus all the other application specific ports required
      by the module. The functional part contains the code that is activated on
      the module:

```
#include "applicationlibrary.h"
SC_MODULE(ATM_test)
{
void watchdog();

SC_CTOR(ATM_test)
{
  SC_CTHREAD(watchdog, clock.pos());
  watching(reset.delayed()==true);
}
};
```

   Analyser.icc file contains the control code and uses the library cells
that are defined in the library, with their access and computational
methods. For example:

```
void ATM_test::watchdog()
{
FIFO<sc_int<16> , s1 > f;

        out.write(0);
        wait();
        while(true) {....}  // functional phase
}
```

2.   Testbench preparation. A test bench is prepared in SystemC-Plus in
      order to simulate and test the specification prior to synthesis. A similar
      test bench is also prepared to be used at SystemC level
3.   The synthesis step is performed, and the file in SystemC generated
      from the specification in SystemC-Plus.
4.   Simulation is performed on the two models (at SystemC-Plus and
      SystemC levels). The two simulations are compared, in order to check
      that the synthesis step did not disrupt the module functionality

5.   The obtained and verified code is given to the commercial synthesis design flow (figure 2), in order to generate lower level descriptions of the design, including VHDL RT.

The testing phase produces traces that can be compared in order to verify consistency of the design flow covering different abstraction levels. Figure 4 shows one such trace, obtained though a free on the market trace.

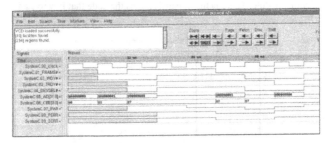

Figure 4. Simulation trace from the SystemC level of description.

A similar methodology is applied to TCP/IP IP applications, i.e. specification of hierarchical FSMs plus proper data-path manipulation of the abstract data types.

6.   CONCLUSIONS

In this paper, we showed how we exploited some of the system design capabilities of SystemC-Plus to model simple telecom cells. The significant benefits are the ease of modeling and the support of reuse. The combined use of specialization hierarchy and templates enables a wide applicability of the library. Another relevant value of the approach has been the possibility to cover two different application domains, like ATM and Internet, with similar *Abstract Data Types*.

7.   REFERENCES

[1]   A. Allara, M. Bombana, P. Cavalloro: Requirements for synthesis oriented modelling in SystemC, Proc. FDL 2001, Lyon, 3-7 September 2001.
[2]   Herrera, Camargo, Villar : Embedded system design methodology based on SystemC, Proc. FDL 2001, Lyon, 3-7 September 2001.

[3]   Bernard, Mueller: SystemC: A case study on behavioural synthesis and simulation performance, Proc. FDL 2001, Lyon, 3-7 September 2001.

[4]   G. Economakos, P. Oikonomacos, I. Panagopoulos, I. Poulakis, G. Papakonstantiou, Behavioral Synthesis with SystemC, Proc. Date'01, 2001.

[5]   R.Pasko, S.Vernalde, P.Schaumont, Techniques to Evolve a C++ Based System Design language, Proc. of Design Automation and Test in Europe, DATE 2002, Paris, France, March 4-8, 2002. pp. 302-309.

[6]   http://www.systemc.org/index.html

[7]   Synopsys, CoWare, Frontier Design, SystemC version 2.0 Users Guide, 2001.

[8]   http://odette.offis.de/systemc-plus/systemc-plus.php3

[9]   E. Grimpe, SystemC object-oriented extensions & synthesis features, Proceedings FDL'2002

[10]   O. Lachish, A. Ziv, Object-oriented High-level Modelling of an InfiniBand to PCI-X Bridge, Proceedings FDL'2002

[11]   ITU-I Recommendation I.432

[12]   http://www.synopsys.com

Chapter 19

A METHOD FOR THE DEVELOPMENT OF COMBINED FLOATING- AND FIXED-POINT SYSTEMC MODELS

Yves Vanderperren {1}, Wim Dehaene {2}, Marc Pauwels {1}
{1} STMicroelectronics Belgium (previously with Alcatel Microelectronics), {2} Katholieke Universiteit Leuven

Abstract: *SystemC is an open source C++ library for creating cycle-accurate or abstract model of hardware architecture, software algorithms and system-level designs. The open nature of SystemC allows users to add features solving issues they faced. Our objective is to propose a solution for the development of joint floating- and fixed-point SystemC models. The presented approach ensures a seamless transition and a flexible solution to analyze the impact of the precision loss along the design.*

1.   INTRODUCTION

SystemC [1] increases the abstraction level of the model of a system, and allows different architectural solutions to be analyzed before the actual implementation as hardware and/or embedded software. In this design and analysis phase, engineers typically start modeling systems using floating-point data types. These handle an almost infinite range of values (depending on the language or the host computer) and simulate fast, but do not reflect the actual implementation.

Within digital hardware, numbers are usually represented as fixed-point data types with a limited precision. Word sizes are fixed at a certain number of bits. We leave out of scope the floating-point hardware implementation. The limited dynamic range of fixed-point values requires careful attention on proper scaling, in order to avoid overflow and/or unreasonable quantization errors. Hence the floating-point model needs to be converted to fixed-point and meet the precision, power and cost constraints of the system. This adaptation entails a significant effort. The designer has to consider the

E. Villar and J. Mermet (eds.), System Specification and Design Languages, 231–242.
© 2003 *Kluwer Academic Publishers.*

numerical range in the system, the required precision, the rounding mode, decide how to deal with exceptional arithmetic conditions. A given fixed-point representation is proposed and the effects of the conversion are verified by means of repeated simulations to get closer to the ultimate target. This time-consuming task introduces an undesired delay in the overall development process of a project.

SystemC provides a fixed-point library extending C++, which has no fixed-point native data types. The SystemC fixed-point data types are accurate to the bit level and support a number of features allowing a high level of modeling. The designer specifies the type of the ports, signals or variables of his SystemC model as either floating-point (e.g. `double`) or fixed-point (e.g. `sc_fixed`). This conversion can be performed fast using an automated approach, as proposed in [2][3] for ANSI-C, but may lead to maintenance issues and error-prone situations since two models exist in parallel, a floating-point and fixed-point version. Any future change of the system needs to be applied to both source codes, or the conversion must be performed again.

Our goal is to achieve a flexible way to develop both floating- and fixed-point SystemC models. The proposed solution enables a single source code to simulate both precisions, which relieves the designer from the conversion activity. Moreover it allows exploring easily the fixed-point design space, e.g. using the optimization method described in [4], without modifying the SystemC code and extra recompilation time.

2.   FLOATING- AND FIXED-POINT VALUES

To clarify the scope, it is useful to review some general information about the representation of fixed-point numbers [5]. In fixed-point hardware, values are stored in binary words, characterized by their word size and the position of the binary point.

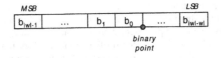

Figure 1. Representation of fixed-point numbers

Figure 1 illustrates the usual representation of a signed or unsigned binary fixed-point number, where

- *wl* is the total word length
- *iwl* the integer word length, or the number of bits on the left of the binary point

The fixed-point value is linked to the unconstrained floating-point number by a slope and bias-encoding scheme

$$n_{fl} \cong n_{fx} = s \cdot n_q + b$$

(1)

where

- n_{fl} is the real-world unconstrained floating point number, i.e. before the conversion of the model from floating-point to fixed-point
- n_{fx} is the real-world fixed-point approximate value
- n_q is an integer value which encodes n_{fx}

$$if\ unsigned:\ n_q = \sum_{i=iwl-wl}^{iwl-1} b_i\, 2^i$$

$$if\ signed:\ n_q = -b_{iwl-1}\, 2^{iwl-1} + \sum_{i=iwl-wl}^{iwl-2} b_i\, 2^i$$

- *s* is the slope; it can be viewed as a scaling factor between n_q and n_{fx}
- *b* is the bias

The slope and the bias do not appear in the hardware, only n_q is visible. Figure 2 illustrates the relationship between these parameters (bias assumed to be zero, and saturation as overflow mode).

3. SYSTEMC ARCHITECTURE

This section summarizes the internal structures of SystemC fixed-point class library [6] in order to show how the solution proposed in this paper fits into it. Figure 3 illustrates the object structure and relations of the main fixed-point class libraries, using the UML notation [7]. There are 4 basic types defined in SystemC: `sc_fix`, `sc_ufix`, `sc_fixed`, and `sc_ufixed`. The fast fixed point types `sc_ufix_fast`, `sc_fix_fast`, `sc_ufixed_fast` and `sc_fixed_fast` are just limited precision versions of the basic types and left out of scope. The types `sc_fixed` and `sc_ufixed` inherit from `sc_fix` and `sc_ufix` respectively, which inherit both from `sc_fxnum`, the base class for the fixed-point types and arbitrary precision.

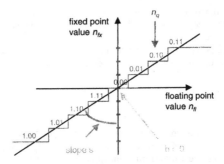

Figure 2. Relationship between floating- and fixed-point representation

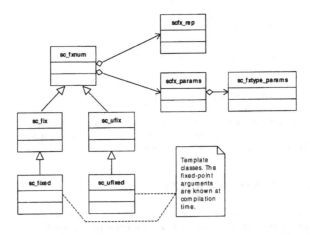

Figure 3. SystemC fixed-point class diagram

Types sc_fixed and sc_ufixed require static arguments to specify the fixed-point characteristics, while sc_fix and sc_ufix types use dynamic arguments. sc_fixed and sc_ufixed are template classes, their static arguments must be known at compilation time and do not change. On the opposite, variables can be used to provide the arguments of sc_fix and sc_ufix objects. Yet the fixed-point type parameters can't change after construction. The complete list of arguments of the SystemC fixed point types is the following:

- wl: the total word length, i.e. the total number of bits used in the type
- iwl: the integer word length, i.e. the number of bits on the left of the binary point (.)
- q_mode: quantization mode, determining the behavior when an operation produces more bits on the LSB side than available
- o_mode: overflow mode, determining the behavior when an operation generates more bits on the MSB side than available
- n_bits: number of saturated bits, used in saturation is selected as overflow mode, to specify how many bits are saturated

In the case of sc_fixed and sc_ufixed, these parameters are provided separately as template arguments; for sc_fix and sc_ufix, an object of type sc_fxtype_params must be declared, initialized and given as an argument in the sc_fix and sc_ufix object declaration. In any case, the designer can query these arguments by means of the public methods provided with sc_fxtype_params.

4.   INTRODUCTION OF THE NEW TYPE

4.1   Implementation

In the SystemC code, the data type of the signals, ports, variables etc. must be specified. Since this information is hard written in the code, it is impossible to have a single model for both floating- and fixed-point precisions. A preprocessor solution, where the type information is replaced either by a floating-point or a fixed-point type, does not benefit from object-oriented programming (OOP) features. It anyway presents many drawbacks, e.g. methods that can't be applied on both types, debugging issues because of discrepancies between original and preprocessed code etc.

The proposed solution (Figure. 4) combines the floating- and fixed-point value by defining new types, called fx_double and ufx_double. They use dynamic arguments and inherit from sc_fix and sc_ufix. The motivation is to avoid the introduction of any parameter that needs to be known at compilation time.

By virtue of inheritance, the new types are fixed-point types and all SystemC fixed-point methods are valid for fx_double and ufx_double. They also carry the floating-point value, hence the SystemC model is able to simulate both precisions without major changes to the source code. The types contain the scale (the slope *s* in eq. (1)), which helps verifying the consistency of the operations performed between two `fx_double` or `ufx_double` numbers: it is e.g. forbidden to add or subtract numbers having different scales. The `fx_double` and `ufx_double` types are type checking and detect a bug that would remain unnoticed with the original SystemC types.

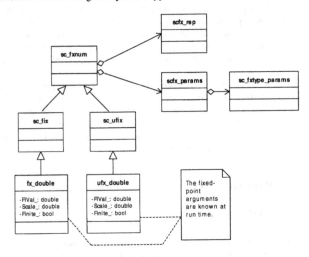

Figure 4. Proposed extension to the SystemC class diagram

Although `fx_double` and `ufx_double` inherit from `sc_fix` and `sc_ufix` respectively, interoperability is ensured as well as cross assignment between signed and unsigned values.

Modules have standardized interfaces using `fx_double` and `ufx_double`, contrary to the template classes `sc_fixed` and `sc_ufixed`. The ports have indeed a unique type that does not involve any template arguments. This increases the reusability of the modules.

There is no need to define methods applicable to each particular statically specified fixed-point type. A single method can be called at different times with fixed-point variables having other characteristics, since these are just dynamic arguments of the type.

Arithmetic operations are computed on both fixed- and floating-point values. Logical operations require the precision mode to be specified, to know on what value they apply. Yet a model performing only computations produces both floating- and fixed-point results in a single simulation run.

A practical case illustrates these principles. A device is measuring water temperature and providing an 8-bit number. Its range is limited between 0 and 100 °C, while the scaling factor is 100/255 °C/bit. The scaling factor can be seen as a measure unit in the conversion between the fixed-point value (n_q) and the floating-point value (n_{fl}).

The following notation is used next:

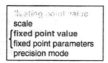

Figure 5. Joint floating- and fixed-point attributes

The example on Figure 6 shows the result of an arithmetic operation applied on two `ufx_double` numbers. The operation is relevant (the slope of the operands are the same), and is computed on both floating- and fixed-point parts. Eq. (1) is verified for the operands but not for the result, as the rounding error propagates. The slope of the result is automatically deduced from the operands.

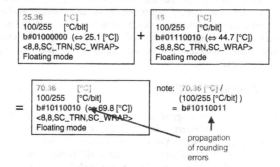

Figure 6. Example of arithmetic operation

Conditional operation (Figure 7) can give different results depending on the precision mode, which must be the same for both operands. Therefore a module having conditional branches still requires separate floating- and fixed-point mode simulation runs, but the SystemC model does not need any conversion. The user specifies the precision mode dynamically at run-time, which saves dramatically compilation time. Moreover a single executable and SystemC model exist.

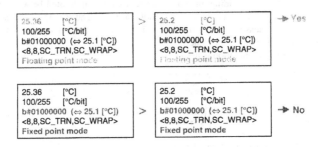

Figure 7. Example of conditional operation

The inherent cyclic property of some quantities map outstandingly onto fixed-point numbers and features. Angles, for example, wrap by construction. This behavior is easily specified by selecting SC_WRAP as the overflow mode when using SystemC fixed-point libraries, but is not applicable for C++ floating-point types. In such cases, it is desirable that floating-point numbers mimic the fixed-point wrapping characteristic, instead of having an infinite range. Therefore an optional parameter of fx_double and ufx_double can be used to let the floating-point part wrap when the maximum value the fixed-point part can represent is reached. This saves extra source code usually required by floating point data types to emulate this feature.

4.2   Cost and benefits

The cost in simulation speed is significant with respect to the SystemC fixed-point types. Experiments show that the scale and precision mode check cause half of the overhead. Yet the added value of such type-checking features is major while designing. Once the model is ready for fixed-point design space exploration, simulations can be speeded up using a version of the single data type without scale and precision check.

Having a single data type for both floating- and fixed-point values enhances the readability of the SystemC code. Module ports have unique types, which increases reusability. Furthermore specific operations do not need to be specified for floating- or fixed-point, saving code development time and reducing risks of discrepancies. A single SystemC model for both precisions prevents maintenance risks in case of bug fixes and during the fixed-point design space exploration. Using the new type, designers are encouraged to think earlier about the fixed-point implementation, avoiding a late conversion from floating-point to fixed-point data types. Automatic scale propagation eases the interpretation of the fixed-point results of arithmetic operations. Dynamic arguments allow changing the fixed-point parameters without extra recompilation and studying easily the impact of these modifications on the design.

5.   EXPERIMENTAL RESULTS

As an example, Figure 8 illustrates the output of a SystemC model computing the autocorrelation of incoming samples, used in the context of a telecommunication application in order to detect the synchronization reference at the physical layer level.

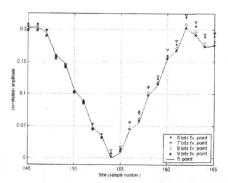

Figure 8. Example of output signal degradation due to fixed-point input and internal signals

The correlation is expressed in terms of the received complex samples $x(n)$ and the correlation length L as

$$\rho_{xx}(n) = \sum_{k=0}^{L-1} x(n-L-k) \cdot x^*(n-k)$$

The horizontal axis represents time (i.e. received sample numbers) and the vertical axis the correlation amplitude $|\rho_{xx}(n)|$. Various fixed-point word lengths of the input samples $x(n)$ have been simulated, in order to measure the accuracy of the correlation output with respect to precision of the samples. The results are saved by the SystemC testbench of the design under test in Matlab file format, in order to avoid writing the numbers by means of output streams [8][9]. This would slow down the simulation, and the precision of the observed results would depend on the format specified when writing to the stream. The simulation results are then analyzed within Matlab in order to quantify and assess e.g. in terms of signal-to-noise ratio (SNR) the impact of the limited fixed-point precision, and find out what is the minimum number of bits that is required to still meet the constraints of the system.

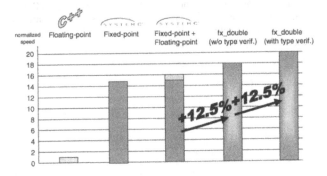

Figure 9. Relative simulation speeds

Figure 9 compares the relative simulation speed[3] of the model when using for $x(n)$ and $\rho_{xx}(n)$ C++ floating-point type, SystemC fixed-point type with static arguments, and the proposed `fx_double`. The first case is taken as a reference. There is actually room for improvement [10], since the autocorrelation model has not been optimized for speed. Moreover a single model is developed instead of two and the design time is reduced, which is not taken into account in Figure 6. Unfortunately this gain is not easy to quantify. The analysis of the fixed-point design space is faster too: the fixed-point word length characteristics of the correlator are provided in a text file and applied dynamically to the model at start of the simulation. Contrary to

---

[3]   Model compiled with g++ 2.95.3 and simulated using SystemC v2.0 on SunOS 5.7 Unix OS.

`sc_fixed`, there is no need to modify the fixed-point parameters in the SystemC model. This saves a significant amount of recompilation time. Typically such a fixed-point design space exploration would require cumbersome and inefficient operations, e.g. manual changes of the model or use of the preprocessor. Moreover this approach would produce different executables, while only one is necessary with the proposed solution. Updated results are available fast if there is any future change to the source code, since the modification is applied to a unique version of the model, not the floating- and the fixed-point versions.

6. CONCLUSION

This paper has introduced a new type that improves significantly the design of combined floating- and fixed-point models, without changing the source code of the SystemC core.

Experiments have confirmed that the presented approach, allowing the development of a joint floating- and fixed-point model, offers multiple advantages and decreases the overall design time. As a conclusion, this could be a powerful extension to the language.

7. REFERENCES

[1] Open SystemC Initiative (OSCI), SystemC v2.0 documentation http://www.systemc.org, 2001

[2] H. Keding, F. Hürtgen, M. Willems, M. Coors, FRIDGE: A Fixed-Point Design And Simulation Environment, DATE 1998

[3] M. Willems, V. Bürsgens, H. Keding, T. Grötker, H. Meyr, System Level Fixed-Point Design Based On An Interpolative Approach, DAC 1997

[4] W. Sung, K. Kum, Simulation-Based Word-Length Optimization Method for Fixed-Point Digital Signal Processing Systems, IEEE Transactions on Signal Processing, vol. 43, pp. 3087-3090, Dec. 1995

[5] The MathWorks, *Fixed-Point Blockset User's Guide v3.0* http://www.mathworks.com, 2001

[6] L. Charest, SystemC v2.0 documentation, http://www.iro.umontreal.ca/~chareslu/systemc-2.0Beta2/, 2001

[7] G. Booch, I. Jacobson, J. Rumbaugh, *The Unified Modeling Language User Guide*, Addison Wesley, 1999

[8] M. Pauwels, A. Berna, F. Ozdemir and Y. Vanderperren, Tutorial: Using SystemC for System-on-Chip Modelling and Design http://www.ittf.no/aktiviteter/2001/dakforum/ program, DAK Forum 2001

[9] Y. Vanderperren, G. Sonck, P. van Oostende, M. Pauwels, W. Dehaene, T. Moore, A Design Methodology for the Development of a Complex System-on-Chip Using UML and Executable System Models, FDL'02 Conference Proceedings
[10]   S. Meyers, *Effective C++* 2nd Edition, Addison Wesley, 1998

Chapter 20

OBJECT-ORIENTED HIGH-LEVEL MODELING OF AN INFINIBAND TO PCI-X BRIDGE

Oded Lachish and Avi Ziv

IBM Haifa Research Laboratory, Haifa University Campus, Haifa 31905, Israel

Abstract: This paper describes a modeling experiment that defines the skeleton for an object-oriented high-level modeling methodology. In the experiment, we built a high-level model for an InfiniBand to PCI-X bridge, using the SystemC class library. Our goal was to examine the initial modeling point and to understand the constructs needed for a model at that level of abstraction.

Key words: High-level modeling, design methodology, object-oriented techniques

1. INTRODUCTION

The rate of advancement in chip manufacturing technology is rapidly outpacing the evolution in design tools and methodology. This induces an ever-increasing gap between the complexity of the design and the ability of designers to grasp the complete behavior required from the design. In the past, this gap caused an evolution in design methodology that raised the level of abstraction in designs from transistors to gates and to *Register Transfer Level* (RTL) design. This change resulted in a rapid increase in designer productivity, which matched the increase in design complexity.

Currently, the highest abstraction level commonly captured by code is RTL. At higher levels of abstraction, a "pen and paper" approach is used. This informal description may contain ambiguities, missing details, and contradictions. Moreover, it is difficult to reason about the design and check its correctness using this method. In *high-level modeling* techniques, the design process starts with a formal executable model of the design, at a level higher than RTL. Then, by either manually refining the model or by using some automatic synthesis tool, the high-level model is converted into a synthesizable model. There are several advantages to this approach. First, it

E. Villar and J. Mermet (eds.), System Specification and Design Languages, 243–253.
© 2003 *Kluwer Academic Publishers.*

allows the design team to concentrate on the real issues and complexities of the design, without worrying about the details required at the RTL level. Second, the design document created by the high-level model is simpler and has fewer ambiguities and contradictions. Finally, the process of verification can begin earlier, before many design decisions are made.

In recent years, an increasing amount of work is being done on high-level modeling issues, including methodology [5], system design [8], and languages [7]. New tools that support this idea are constantly being developed [2,3]. Yet, only a small number of actual products are designed using high-level modeling. Many claim that the reason for the low utilization of high-level modeling tools is that designers are conservative, and therefore reluctant to use new tools. Another reason is the high cost of using high-level modeling, either at design time, or in chip parameters. Another compelling reason for the low popularity of high-level modeling techniques is the absence of a solid design methodology based on high-level modeling.

This paper describes a modeling experiment in which we built a high-level model for an InfiniBand [10] to PCI-X [6] bridge. The model was built as part of an experiment, whose goal was to define a skeleton for an object-oriented high-level modeling methodology for ASIC design. The experiment was conducted specifically to examine the initial modeling point and understand the building blocks needed for a model at that level of abstraction. Other goals were to investigate the use of object-oriented techniques in the model and the incorporation of cores in high-level models. This experiment is part of the ODETTE project [12], whose goal is to investigate the use of object-oriented techniques in hardware design.

The first major issue in designing a bridge between two protocols is the interaction between the protocols. Therefore, we chose to model at a level that captures these interactions. Our model is strongly based on the object-oriented programming paradigm provided by C++ and takes advantage of some of its characteristics, including encapsulation, and inheritance. Hardware constructs, such as signals, are represented in our model using SystemC [11] and its object-oriented extension, SystemC-Plus [1]. The construction of the model is based on four types of objects: data, structural, functional, and algorithmic. These different types allow us to easily capture different aspects of the behavior of the model. The model we created helped us understand the interactions between the InfiniBand network and the PCI bus. We could easily simulate many scenarios involving such interactions and verify that our design decisions resulted in the expected behavior.

The rest of the paper is organized as follows. In Section 2, we describe the specification of the InfiniBand to PCI-X bridge. In Section 3, we discuss some of the modeling issues faced during the experiment. Section 4 describes the construction of the model of the bridge.

2.   INFINIBAND TO PCI BRIDGE SPECIFICATION

The model we built is a bridge between the InfiniBand network [10] and a PCI-X bus [6]. The main goal of the chip is to remotely connect I/O devices on a PCI-X bus to host CPUs over the InfiniBand network. The bridge for the InfiniBand network is a *Target Channel Adapter* (TCA), an adaptor that connects between the InfiniBand network and I/O devices. On the PCI-X bus side, the bridge emulates the behavior of a local host.

InfiniBand is a layered architecture over a switched I/O. The bridge is specified to implement three out of the four lower layers of the InfiniBand architecture: the physical layer, which is responsible for the physical connection; the link layer, which is responsible for packet transfer between neighboring elements; and the transport layer, which is responsible for end-to-end transfer of messages.

Figure 1 shows a functional description of the bridge and highlights the layered structure of the InfiniBand architecture. The main part of the chip is the transport layer of the InfiniBand network, which handles the transport protocol and the conversion to and from PCI-X transactions.

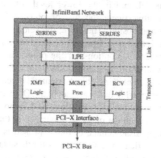

Figure 1. Functional description of the InfiniBand to PCI-X bridge

The InfiniBand transfer protocol supports several modes of data transfer. The specification of the bridge uses only the *reliable connection* mode that handles reliable point-to-point transfer between two network adaptors. The main components in the transport layer are *queue pairs* (QPs). Each QP handles a single InfiniBand connection, and consists of receive and transmit queues. The receive queue receives InfiniBand packets, assembles them into messages, and converts them to the appropriate PCI-X transactions. The transmit queue receives PCI-X transactions, converts them into InfiniBand messages, and sends them to the InfiniBand network. Routers and arbiters are responsible for directing incoming traffic to the proper QP and selecting the next packet/transaction to be handled.

3. HIGH-LEVEL MODELING ISSUES

Several issues must be addressed when building a methodology for high-level modeling. These issues include the abstraction levels for the start and end points of the methodology, the design paradigm and programming language most suitable for the implementation of the models, and the computational model that should be used. In this section, we discuss some of the issues addressed in our modeling experiment.

3.1 Design Paradigm and Programming Language

The first issue we addressed was the selection of the programming language and design paradigm for the implementation of the model. Our first requirement was that the programming language be appropriate for use both in the high-level model, and in the lower levels of descriptions. Another requirement was to investigate the use of object-oriented techniques in hardware modeling and design. Therefore, the selected language needed to support the object-oriented paradigm on one hand, and hardware description constructs on the other. Our options for the programming language of the model were to add object-oriented constructs to an existing HDL, or the add hardware constructs to an existing object-oriented language.

Several object-oriented extensions such as Objective-VHDL [4] and Superlog, have been defined for existing HDLs, however, their use is minimal. The major advantage of these languages is their strong hardware foundation. The major disadvantage of this approach is the lack of tools to support debugging, verification, and synthesis for these languages.

There are number of languages that extend C++ with hardware constructs, such as concurrent behavior, signals, and modules. SpecC [9] and SystemC [11] present two possible approaches for defining such languages. SpecC defines a hardware design language on top of C++ by adding new keywords that support parallel behavior and structural description. This requires an extension or even a redefinition of the language and a dedicated compiler. SystemC adds hardware constructs to C++ in the form of a class library. Consequently, any future expansions only require additional libraries and simulation of such a language can be done using a standard compiler.

The SystemC approach was the most suitable for our needs and met all our requirements. Moreover, the flexible class library approach allowed us to use *SystemC-Plus* [1], an object-oriented enhancement of SystemC. One example of the SystemC-Plus enhancements is the ability to transfer objects of derived classes in signals of a base class.

3.2   Programming Methodology

One of the features of the C++ programming language is the freedom of expression it gives to users. While this freedom can lead to creative programming, it can also lead to complex programs and programs that are hard to refine into hardware designs. To maximize the reusability of the model and increase our ability to refine it, we restricted ourselves to four types of classes that capture different aspects of the model. The four class types used are data classes, structural classes, functional classes, and algorithmic classes.

Data classes are used to encapsulate the data elements that flow in the design. These classes are the basis for high-level protocols and data flow. Each data class contains a set of private variables and a set of methods over the private variables. Data classes have several advantages over the use of raw data. With data objects, we do not have to deal with the format of the data. Moreover, data classes allow us to postpone decisions on the exact implementation of some of the functionality related to data objects, to later stages in the design. Another advantage of data classes over raw data is the ability to piggyback additional information that is needed for debugging, verification, and performance evaluation. For example, adding routing information to a packet object can help verify that the packet followed the correct route and identify bottlenecks.

The other three types of classes represent three aspects of the broader functionality of the design. *Structural classes* capture structural aspects of the design. They are implemented as SystemC modules, with functionality that is implemented in synchronous and asynchronous processes. Structural classes use signals to communicate with each other and with the outside world, and method calls to communicate with classes of other types.

Functional classes represent pure functionality without structure. They can contain instances of other functional classes and data classes. Examples of functional classes are storage elements such as FIFOs. The functional objects should be as autonomous as possible, to allow reusability.

Algorithmic classes are a special case of functional classes. They implement a specific algorithm without an internal state. We found this type of class very useful in reducing the interdependencies between functional classes. For example, we used them to connect classes related to the InfiniBand network with classes related to the PCI-X bus, while maintaining the generic nature of the functional classes to promote reuse.

3.3   Choosing the Proper Level of Abstraction

The abstraction level used for modeling should be chosen as a direct result of the goals of the model. For example, a model that explores schemes for data sharing is different from a model that examines interaction between two protocols. The complexity of the specification and the designer's experience may also influence the decision regarding the level of abstraction.

The InfiniBand to PCI-X bridge that was the target of the experiment is one of the first InfiniBand chips designed at the IBM Haifa site. Therefore, one of the main goals of the experiment was to model the behavior of the InfiniBand protocol and its interaction with the PCI-X protocol. This led to a very high level of abstraction in the model, with most of the functionality of the model described in functional and algorithmic classes. Once we chose the level of abstraction, we had to decide which concepts capture this level of abstraction, namely, representation of data, structure, functionality, and time granularity. These decisions are described further in Section 4.

3.4   Using Preexisting Cores

The use of preexisting cores in ASIC design is rapidly gaining popularity. Therefore, any high-level modeling methodology must address this aspect of the design. Correct handling of preexisting cores is especially important when high-level models of these cores do not exist. There are several possible ways to handle cores when high-level models are not available. The first option is to use a low-level model of the core. This has the advantage of giving us an exact model of the core. Nevertheless, the low-level model can slow down simulation. Moreover, the low-level core may force us to prematurely use an exact interface in the high-level model.

A second option is to replace the core with a stub. This option is useful when the core plays a small role in normal operation of the design. For example, in our model, we replaced the management processor with a stub, because the management processor had minimal involvement in the normal operation of the bridge. Another option is to incorporate the functionality of the core in the rest of the model. This option is useful when small parts of the core functionality are needed. In our case, small parts of the functionality of the link layer were needed in our abstract model. Therefore, we decided to incorporate the relevant functionality of the link layer core in the transport layer unit. Finally, if all other options are not suitable, we can create a high-level model of the core. While this option may require a lot of effort, it can pay extra dividends when a high-level model of the core is reused in future designs. The effort required to create such a model may also be reasonable because of the high-level of abstraction used in the model.

4. MODEL CONSTRUCTION

We constructed our model to reflect the basic decisions usually made at the start of the chip design process, as well as the modeling issues we discussed in the previous section. Therefore, the first two issues we addressed were the abstraction level for the model and the partitioning of the bridge. Once we made these decisions, we could begin constructing the part of the model functionality that was not assigned to preexisting cores.

4.1 Partitioning

We partitioned the model into five major units, based on common concepts used to partition chips of this type. The partitions of the model are illustrated in Figure 1. The partitions include units for the PCI-X bus interface, the management processor, and each of the three layers of the InfiniBand protocol implemented in the bridge.

The management processor, whose operation has little influence on the normal operation of the bridge, is implemented as a simple stub in the model. In the link layer core, only the validity checks of incoming packets are needed. These checks are incorporated into the InfiniBand packet classes. The PCI-X bus interface core is responsible for the initiation and completion of outgoing transactions, and the identification and reception of incoming transactions. Because the operation of this unit is important to the end-to-end operation of the bridge, we implemented a simple high-level model of this core in an abstraction level that matched the rest of our model.

The InfiniBand transport layer unit supports the functionality of the InfiniBand transport layer as required in the bridge and the management of the interaction between the InfiniBand network and the PCI-X bus. This unit is the only unit that is not implemented in the chip as a preexisting core.

4.2   Realization of the Chosen Abstraction Level

The abstraction level used in the model is designed to capture the interaction between the InfiniBand network and the PCI-X bus, and the basic structure of the final design. The abstraction level influenced many aspects in the structure and content of the model, and led to specific representations of data, structure, and functionality. The first aspect is representation of InfiniBand packets and PCI-X transactions. At the abstraction level we used, we were not interested in the exact representation of the data and the exact mechanisms used to transfer data. Therefore, we found it best to encapsulate the InfiniBand packets and the data transferred on the PCI bus in data objects. This representation level helped us avoid dealing with some of their functionality. For example, we replaced the calculation of the CRC in InfiniBand packets with a predetermined result of the CRC check.

For simplicity, our model was synchronous. We assumed that sending or receiving a single packet and a single PCI transaction takes one cycle. The coarse timing granularity allowed us to focus on the interactions between InfiniBand and PCI, which was the goal of the model. On the other hand, it didn't allow us to check issues of resource sharing in the model.

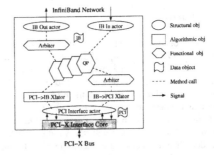

Figure 2. Main classes in the model

Figure 2 shows the main classes used in our model, and lists the type of each class. The most natural place to identify the data classes in the model is at the interfaces. Therefore, our basic data classes are InfiniBand packets and PCI transactions. The InfiniBand packet classes encapsulate portions of the bridge functionality. For example, some of the validity checks of packets are implemented as methods in our InfiniBand packet classes. On the PCI bus, data transfer is done via a continuous handshake between the master and the slave. Therefore, PCI transaction objects contain both the request of the master and the response of the slave.

The data classes in the model are the classes that use the object-oriented features of C++ the most. In addition to the encapsulation described above, these classes also use inheritance and polymorphism to handle different types of InfiniBand packets and PCI transactions. Using different classes for different types of packets, helped to simplify the code of the data classes and reduce development and maintenance effort.

The structural objects form the skeleton of the chip's structure. Because we wanted to limit the structural elements in our model, we used them only to implement the interfaces of the chip. We used two modules for the input and output of the InfiniBand network, along with an additional module that handles the interface between the transport layer and the PCI interface core.

The functional classes implement most of the functionality of the bridge. The classes in our model closely follow elements defined in the InfiniBand protocol. Consequently, the two major functional classes we identified are queue pairs (QP) and arbiters. The QP class contains information about the connection that uses the queue pair, methods that check the validity of incoming packets, and methods that handle acknowledgments.

In addition to queue pair classes, we implemented two types of arbitration classes: one required by the InfiniBand architecture to select the next packet to transmit, and one that selects the next PCI transaction to be initiated by the bridge. Because the functionality of both classes is similar, they inherit most of their functionality from a generic arbiter class.

Both the queue pair and the arbiter classes are generic and can be used in the transport layer of any InfiniBand chip. The specific functionality of the bridge in the transport layer is implemented in two algorithmic classes that convert between InfiniBand messages and PCI transactions.

4.3 Model Statistics

Table 1 provides statistics about the code in the classes of the model. The table shows the number of classes and lines of code (LOC) for each of the four class types described above. The table shows that the data classes comprise about half of the classes and a third of the code in the model. Modeling each type of InfiniBand packet and PCI transaction as a separate class led to the large number of data classes, while the functionality implemented in these classes contributed the relatively large amount of code. Our attempt to emphasize the functionality of the chip meant that structural classes played a limited role in the model. Therefore, these classes contain less than 15% of the code in the model.

Table 20-1. Model Statistics

Class Type	Classes	LOC
Data	16	1550
Functional	6	1400
Algorithmic	2	600
Structural	4	600
PCI Core	2	550
Total	**30**	**4700**

The effort for the development of the model was roughly four person months. This effort is much too high for the model to be included as part of a design flow. However, this effort included the development of the methodology and several failures. We believe that a similar model can be developed in about a month during the specification and high-level design phase of the development process. Given the benefits of the model, this effort can be worth spending in many cases.

5.   CONCLUSIONS

This paper describes the construction of a high-level model of an InfiniBand to PCI bridge. We modeled the chip using the object-oriented paradigm of C++, with the addition of the SystemC class library to support concurrent behavior and other hardware constructs. The model was created using a simple programming methodology that is based on four types of classes, namely, data, structural, functional, and algorithmic classes.

Building the high-level model for the bridge helped us understand the interactions between InfiniBand and PCI-X. The model made it easier for us to simulate the different protocols supported by the bridge. Yet, the abstraction level we chose did not capture some important design issues, such as concurrent access to shared resources. Therefore, the abstract model we created needed to be followed by a lower level cycle-accurate model that would allow designers to address these issues.

6.   REFERENCES

1. E. Grimpe and F. Oppenheimer. Aspects of object-oriented hardware modeling with SystemC-Plus. In *Proceedings of the 4th Forum on Design Languages*, 2001.
2. T. Kuhn, T. Oppold, C. Schulz-Key, M. Winterholer, W. Rosenstiel, M. Edwards, and Y. Kashai. Object oriented hardware synthesis and verification. In *the 14th International Symposium on System Synthesis in Canada*, 2001.
3. A. Mycroft and R. Sharp. The FLaSH project: Resource-aware synthesis of declarative specifications. In *Proceedings of the International Workshop on Logic Synthesis*, 2000.

4. M. Radctzki. Overview of Objective VHDL language. In *the Proceedings of the 2nd Forum on Design Languages*, pages 279-291, 1999.
5. R. Roth and D. Ramanathan. A high-level hardware design methodology using C++. In *4th High Level Design Validation and Test Workshop*, pages 73-80, 1999.
6. E. Solari. *PCI and PCI-X Hardware and Software*. The Coriolis Group, 2000.
7. D. Verkest, J. Cockx, F. Potargent, G. de Jong, and H. De Man. On the use of C++ for system-on-chip design. In *Proceedings of the IEEE Workshop on VLSI*, pages 42-47, April 1999.
8. D. Verkest, J. Kunkel, and F. Schirrmeister. System level design using C++. In *Proceedings of Design, Automation and Test in Europe*, pages 74-83, 2000.
9. J. Zhu, R. Doemer, and D. D. Gajski. Syntax and semantics of the SpecC language. In *the Proceedings of Synthesis and System Integration of Mixed Technologies*, 1997.
10. InfiniBand architecture, general specifications, 2000.
11. SystemC 2.0 user's guide. http://www.SystemC.org/.
12. ODETTE - Object-oriented co-DEsign and functional Test TEchniques. RTD project IST-1999-11476 of the European Commission. http://www.odette.offis.de.

Chapter 21

A SYSTEMC MODEL FOR RTOS KERNEL

Mohamed Abdel Salam, Ashraf Salem
Mentor Graphics Egypt, Cairo, Egypt

Abstract: This paper presents a new methodology for modeling a priority-based preemptive real time operating system (RTOS) kernel in SystemC. We use the current modeling constructs of SystemC 2.0 and throughout our development of the kernel's system calls; new constructs dedicated to RTOS modeling have been proposed. The article focuses on kernel architecture and shows its interaction with hardware module representing a bus functional model (BFM) of a generic microcontroller. The usage of the proposed kernel's APIs is demonstrated by an embedded software example.

Key words: SystemC, RTOS, Hardware / Software Co-Verification

1.   INTRODUCTION

SystemC provides great flexibility in modeling, for both hardware and software of an entire system throughout the design flow [1][2]. It fits smoothly within current System on Chip (SoC) design methodologies [3] and receives great momentum from both EDA factory and field towards its standardization. The language is entirely based on C++ and its constructs are provided as a C++ class library, for modeling at different levels of abstraction from system behavioral to Register Transfer Level (RTL). In addition, it embeds a lightweight simulation kernel for cycle-based simulation [1][4]. In SystemC 1.0, modeling constructs address RTL to play the same role as VHDL and Verilog in modeling hardware. In SystemC 2.0, new constructs were added to give the language an advantage in describing systems at higher level of abstraction. Current development is to extend SystemC capability to describe embedded software & model RTOS

E. Villar and J. Mermet (eds.), System Specification and Design Languages, 255–264.
© 2003 *Kluwer Academic Publishers.*

[1][4][5]. Co-design [3][6] and Co-verification [7][8] can then be easily achieved using the same modeling language in a homogenous development environment where mature C++ development tools are used to debug various parts of design and then use a process of gradual refinement for synthesizing directly from C/C++ specification [9][10]. In this paper a new approach to model an abstract RTOS kernel using SystemC is proposed. The model describes RTOS architecture, system calls and interaction with a hardware module representing a BFM and a group of associated hardware peripherals. The rest of this paper is organized as follows. In Section 2, we present an overall picture of the model architecture. In Section 3, we highlight some aspects in kernel internal implementation, data structure and algorithms. In Section 4, we show how the software module will be interacting with the hardware module. In Section 5, we present an example showing how to write an embedded software application on the top of the kernel model. Conclusion is in Section 6.

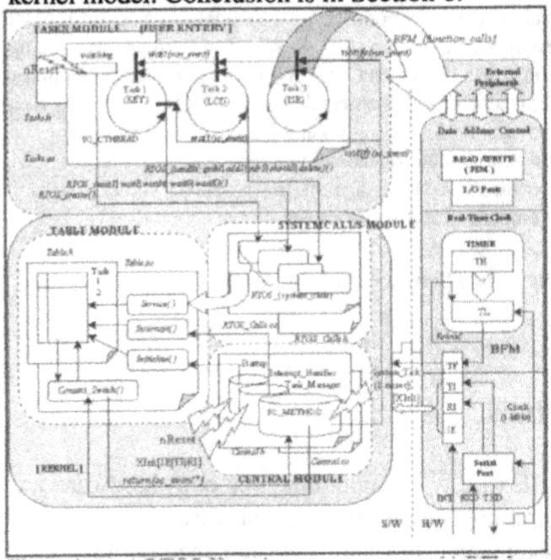

Figure 1:   a) RTOS Kernel   b) BFM

2.   RTOS KERNEL ARCHITECTURE

RTOS kernel architecture is modeled using four interactive SystemC modules (Figure 1.a): Central, Table, System Calls and Tasks modules. We support the following features for each task in kernel model: waiting for an interval event, waiting for a timeout event, handling Interrupts, message & semaphore events.

The proposed model uses the timing feature of a bus functional model (BFM) of a generic microcontroller [13] developed in SystemC (Figure 2), where one timer/interrupt (Figure 1.b) is used to drive the kernel operation with a default timing resolution 1 msec and the capability of adjusting it upon startup.

```
#define TH_COUNT 250
/* When Simulation Clock is 1 MHz & TH = 250 */

#define ROLL_OVER 4   /* 4 Count Rolls give 1 msec Tick */

  SC_MODULE(RT_Clock){            /* Ports */
    sc_out< bool > System_Tick;    /* System Tick */
    sc_in_clk       Clk;            /* Simulation Clock */
    sc_in< bool >  nReset;           /* System Reset */

    void do_RT_Clock();

    unsigned int Roll_Over;        /* Roll Over Counter */
    sc_uint<8> TH, TL;             /* Timer Registers */
    bool TF;             /* Timer Interrupt Flag */

    SC_CTOR(RT_Clock) {
      SC_CTHREAD(do_RT_Clock, Clk.pos());
      TH = TH_COUNT; TL = TH_COUNT; TF = 0; Roll_Over = 0; }
  };
```

Figure 2: Real Time Clock Module

The central module (Figure 3) consists of three SC_METHODs and responsible for tasks' management/switching, kernel's table initialization and interrupt handling. The three methods are: Task manager, Interrupt Handler and startup method. Task manager method is synchronized with every System Tick to call a *helper function* in Table Module to find out which Task to notify next, based on assumed rules of priority and preemption.

```
if (System_Tick.read() == 1){
                System_Tick.write(0);
                notify_event = Context_Switch(Hash_TB);
                if(notify_event)
                notify(*notify_event);

    }
```

Interrupt handler method is sensitive to interrupts (external or Serial). Upon receiving an interrupt, it identifies its type and call function in Table Module to find out which Task to notify in order to handle that interrupt. A hash table HashTB is used to implement the Kernel Table structure. A group of functions are used to manipulate with the hash table such as getHash, putHash and hashIndex.

```
// Interrupt Handler Method
        if(eXInt.read()){    /* External Interrupt */
           eXInt.write(0);
           notify_event = Interrupt(Hash_TB, eXt_INT);
           if(notify_event)
             notify(*notify_event);

    }
```

```
SC_MODULE(Central){                    /* Ports */
   sc_inout< bool >   System_Tick;   /* System Tick */
   sc_inout< bool >   eXInt;          /* External Interrupt */
   sc_inout< bool >   TxInt;            /* Serial Tx Interrupt */
   sc_inout< bool >   RxInt;          /* Serial Rx Interrupt */
   sc_in_clk          Clk;   /* Simulation Clock */
   sc_in< bool >      nReset;           /* System Reset */

   /* Task Management - Context Switching */
   void Task_Manager();

   /* Identifying Type of interrupt and Notifying ISR */
   void Interrupt_Handler();
   void Startup();     /* Kernel Table initialization */

   SC_CTOR(Central) {
     SC_METHOD(Task_Manager);
     sensitive_pos << System_Tick << Clk;
     SC_METHOD(Interrupt_Handler);
     sensitive_pos << RxInt << eXInt << TxInt << Clk;
     SC_METHOD(Startup);
     sensitive_pos << nReset << Clk;
   }
};
```

Figure 3: RTOS Central Module

The startup method is sensitive to active low reset and it is used to initialize the kernel table structure.

```
// Startup method
        if (nReset.read() == 0){
             if(!TABLE_INIT){
                 Hash_TB = Initialize();
                 TABLE_INIT = 1;}}
```

System Calls module (Figure 4) contains all kernel related system calls. They are invoked from a running task, to perform an RTOS service. Each RTOS Call will invoke its equivalent function in the Table Module to update the kernel table structure based on the service needed. *RTOS_wait* uses SystemC *wait(sc_event)* to cause a suspension of the task's thread of execution until its event arrives. RTOS tasks are created using RTOS_create and RTOS_delete. Different wait system calls implements RTOS tasks suspensions: RTOS_waitT implements waiting timeout event. RTOS_WaitI implements waiting Interval events. RTOS_waitX implements waiting external interrupts. RTOS_waitM models task waiting for a message arrival. Semaphores manipulation is done via three system calls, RTOS_addS, RTOS_subS and RTOS_checks. The first two acts as free and lock. The third checks the value of the semaphore. RTOS_getM and RTOS_sendM model messages sending and receiving between tasks.

```
RTOS_waitT: Sleep Until Timeout Event
void RTOS_waitT(char *Task_IdNo, unsigned int Timeout,
    sc_event *Timeout_event){
        Service_RTOS_waitT(Hash_TB, Task_IdNo, Timeout, Timeout_event);
        wait(*Timeout_event);}

RTOS_waitI: Sleep Until Interval Event
RTOS_waitM: Sleep Until a Message arrives in a MailBox
RTOS_waitX: Sleep Until external Interrupt arrives
RTOS_waitS: Sleep Until Semaphore counts zero

RTOS_getM: Fetch a Message when it arrives
char* RTOS_getM(char *Task_IdNo)
  {return(Service_RTOS_getM(Hash_TB, Task_IdNo));}

RTOS_sendM: Send a Message to another Task.
RTOS_addS: Add units to Task's Semaphore.
RTOS_subS: Subtract number of units from a Task's Semaphore
RTOS_checkS: Return a Task's Semaphore Count.
RTOS_create: Register Task in Kernel Table
RTOS_delete: Un-register Task from Kernel Table.
```

Figure 4: RTOS System Calls

3.   TABLE MODULE

Table Module is a hash table that keeps a record on every software task created upon startup and gets updated every System Tick, interrupt and system call service handled. Task information is modeled using a structure called TASKINFO (Figure 5), which contains task id, task status, task type, message, semaphores and Interrupts data. The hash table is updated through a group of functions: *Context_Switch(), Interrupt(), Service() and Initialize()*. Initialize is called from *Central::Startup()* upon receiving a system reset to create the kernel table. Eventually, this function will call *makeHashTable()* to allocate enough memory. For every system call defined in System Calls Module, there exists an equivalent service function to update the task's info structure in the kernel table based on the service required.

For example the function Service_RTOS_waitT updates the task information when RTOS_waitT system call is invoked, by inserting timeout value and timeout_event and mark the task status as ASLEEP.

```
int Service_RTOS_waitT( hashtable *Hash_TB, char *Task_IdNo,
                        unsigned int Timeout,sc_event *Timeout_event);
```

```
enum State {ASLEEP, READY};
enum Int_Type {APPL_TASK, SERIAL_INT, eXt_INT};
typedef struct TASKINFO {
char Task_IdNo; /* Task Id No */
unsigned int Priority; /* Priority */
State Task_S; /* Task State: READY or ASLEEP */
unsigned int Msgbx_F; /* Message box Flag */
char *Msg; /* Message */
unsigned int Timeout; /* Wake me when this time expires
*/
unsigned int Interval; /* Wake me every scheduled time
*/
unsigned int Semaphore; /* Counting Semaphore */
unsigned int Interrupt_F; /* eXt Int Flag */

  /* Type of Task: Application, ISR-eXt, ISR-Serial */
Int_Type Int_T;
sc_event* Msg_event;                  /* Message arrives */
sc_event* Semaphore_event;  /* Semaphore reaches 0 */
sc_event* Timeout_event;        /* Timeout expires */
sc_event* Interval_event;           /* Interval over */
sc_event* Interrupt_event;     /* external interrupt */
sc_event* Run_event; /* If I'm still not preempted */
} Task, TaskPtr;
```

Figure 5: Taskinfo data structure

Context Switch is called from *Central::Task_Manager* METHOD every System Tick to update tasks' timing info in the kernel table & determines which task to run next based on priority and preemption rules. Interrupt function *is* called from *Central: Interrupt_Handler* METHOD upon receiving an interrupt to determine which task to handle the received interrupt.

4.   TASK MODULE AND BFM INTERACTION

Task module represents the programmable entry for the user to implement his tasks. It consists of two types of threads, one that simply represents an application software task for example writing to an LCD or scanning a keypad, another dedicated to an interrupt service routine (ISR) handling one type of interrupts. An application software task can be in one of 3 states: RUNNING, ASLEEP or READY. Within each task, the user is able to access functions located in System Calls Module for example sending a message to another task or waiting for a timeout event, as well as, access member functions of the BFM to interact with hardware peripherals for ex, reading from an external memory location or writing to an I/O port. We used events and dynamic sensitivity introduced in SystemC 2.0 to implement preemption feature in the kernel, where each thread is suspended every cycle of execution and then dynamically notified from the Central Module with every System Tick to keep on running if it's not preempted or go to the READY state if it's preempted, as shown in Figure 6.

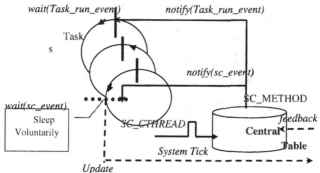

Figure 6: Task's Management

We also used global watching feature of SystemC to watch a system reset calling a system call to create and initialize the software tasks; i.e. register each task and set its priority and initial state. In our methodology, the embedded software is compiled for the host other than the intended CPU[14]. As the running tasks execute, they are able to call member functions of the BFM using an interface based on driver model of communication with an underlying handshaking protocol [15].

5. EMBEDDED SOFTWARE EXAMPLE

A simple example [16] has been implemented, showing two application tasks and two interrupt service routines working together on the top of the RTOS kernel and interacting with the BFM and its associated hardware peripherals. In the following SystemC model, two application tasks are shown. A high priority Task 1 (Figure 7), acts as a scheduler that's executed every 100 msec to: Scan a Switch module connected to I/O Port 1, display scanned data on a LED module connected to I/O Port 2 and send a message to Task 2. A Low priority Task 2 is waiting for a message to arrive from Task 1 to: Display the data on an SSD counter connected to I/O Port 4 and save the most recent value of data in an external data memory at location 0x2010.

```
/* Tasks.cc */
...
void Tasks::Task1() {
 sc_event Interval_event;
                         /* S/W Reset Action */
if(nReset.read() == false)
RTOS_create("Task_1", 0, &Task1_run_event, APPL_TASK);
                         /* A High Priority Task */

while (true){
        /* Suspend until dynamically notified */
    wait(Task1_run_event);

    /* Scheduler: 100 msec */
    RTOS_waitI("Task_1", 100, &Interval_event);

    BFM_RdPort(0x1);        /* Port 1 - Switch */
    while(BFM_PortBusy())
     wait();    /* Pass Control To H/W Thread */
    Test_Data = BFM_FetchPortData();

    BFM_WrPort(0x2, Test_Data);/* Port 2 - LED */
    while(BFM_PortBusy())
     wait();
                /* Send a Message To Task 2 */
    RTOS_sendM("Task_2", "Count");
    }
}
```

Figure 7: Embedded Software Example

In Figure 8, an Interrupt Service Routines (ISR) is shown. Task 3 handles an external interrupt. Upon receiving the interrupt it drives a stepper motor connected to I/O Port 3 to move in a single step fashion. Task 4 handles serial port interrupts. Initially, a transmit interrupt flag is set. This calls a transmit interrupt handler to send a prompt character '>' to a serial buffer. A receive interrupt handler then dumps characters received from the serial buffer to stdout. The peripheral models set {Switch, LCD, Seven-Segment Display counter, DAC monitor, Keypad, Memory Dump, Emulation for a Terminal, Stepper motor and an interrupting device}, are wrapped in GTK+ GUI widgets to give the look and feel of a virtual system prototype, where the user can interact with peripheral devices and visualize the results of running tasks. A VCD waveform viewer is used to monitor memory cycles & I/O transactions. We implemented and tested all our models under Linux RedHat 7.0 using GNU C++ Compiler version 2.95.2 & GNU Make.

```
void Tasks::Task4()
{          /* Serial Interrupt Service Routine */
    unsigned int Char_Rx = 0x00;

        /* Interrupt Initialization */
    IntReg    intreg;
    IntRegPtr intreg_Ptr;

    intreg.EA  = 1;                      /* Enable Global Interrupt    */
    intreg.ES  = 1;                      /* Enable Serial Interrupt    */
    intreg.TI  = 1;                      /* Cause Interrupt to Start   */
    intreg.REN = 1;                            /* Enable Receiver */
    BFM_WrIntReg(&intreg);          /* Initialize Interrupt Register */

    /* Task 4 Reset Action */
    if(nReset.read() == false)
        RTOS_create("Task_4", 3, &Task4_run_event, SERIAL_INT);

    while(true)
    {
        wait(Task4_run_event);

        intreg_Ptr = BFM_RdIntReg();
        if(intreg_Ptr->TI)
        {
            BFM_WrSBUF(0x3E);                       /* Send Character '>' */
            while(BFM_SerialBusy())
                wait();                    /* Pass Control To H/W Thread */
            intreg_Ptr = BFM_RdIntReg();
            intreg_Ptr->TI = 0;
        }
    if(intreg_Ptr->RI)
        {
            Char_Rx = BFM_RdSBUF();                   /* Read Character */
            while(BFM_SerialBusy())
                wait();                    /* Pass Control To H/W Thread */
            intreg_Ptr->RI = 0;
    printf("Rx: Character Read ... %c - 0x%x\n", Char_Rx, Char_Rx);
        }
    }
}
```

Figure (8): Interrupt Service routine

6.   CONCLUSION

   In this paper, we focused mainly on presenting the algorithms and guidelines we propose to model the software side of a complete system architecture representing a priority based preemptive RTOS kernel and a BFM of a generic microcontroller together with a group of associated hardware peripherals using SystemC. Using SystemC to model all parts of the system provided us with great flexibility in design where all features and development tools of standard C++ are used. Using direct C++ calls in the interface design between the software and hardware showed us faster Co-simulation and the performance was very suitable for Co-verification at this high level of abstraction. Finally, it was obvious to us that by proper modeling of the RTOS and BFM, performance estimates can be obtained for software execution, which can be used to drive the Co-design process and speedup system integration phase early in design cycle.

7.   REFERENCES

[1] SystemC User's Guide Version 1.1, Functional Specification Version 2.0, Sept. 2001, http://www.SystemC.org

[2] S. Y. Liao, "Towards a new standard for system-level design," in Proc. 5th International Workshop on Hardware/Software Co-Design, May 2000.

[3] W. H. Wolf, "Hardware/Software Co-Design of Embedded Systems," in Proc. IEEE, July 94.

[4] S. Liao, S. Tjiang, R. Gupta, "An Efficient Implementation of Reactivity for Modeling Hardware in the SCENIC Design Environment," in Proc. DAC'97, June 1997.

[5] J. Gerlach, W. Rosenstiel, "System Level Design Using the SystemC Modeling Platform," Worshop on System Design Automation, SDA 2000.

[6] D. E. Thomas, J.K. Adams, H. Schmit, "A Model and Methodology for Hardware-Software Co-Design," in Proc. IEEE Design and Test, Sept. 1993.

[7] L. Séméria, A. Ghosh, "Methodology for Hardware/Software Co-verification in C/C++," in Proc. ASP-DAC, 2000.

[8] D. Harris, D. Stokes, R. Klein, "Executing an RTOS on Simulated Hardware using Co-verification," in Proc. Embedded Systems Conference, San Jose, Sept. 2000.

[9] G. De Micheli, "Hardware Synthesis from C/C++ Models," in Proc. DATE'99, March 1999.

[10] A. Ghosh, J. Kunkel, S. Liao, "Hardware Synthesis from C/C++," in Proc. DATE'99.

[11] T. W. Schultz, "C and the 8051 Building efficient applications, Vol. II", Prentice-Hall, 1999.

[12] Ward, P., S. Mellor, "Structured Development for Real-Time Systems", Prentice Hall, 1985.

[13] T. W. Schultz, "C and the 8051 Hardware modular programming and multitasking, Vol. I", Prentice-Hall, 1998.

[14] B. Bailey, R. Klein, S. Leef, "Hardware-Software Co-Simulation Strategies for the Future," Mentor Graphics Co., http://www.mentor.com

[15] K. Svarstad, G. Nicolescu, A. Jerraya, "A Model for Describing Communication between Aggregate Objects in the Specification and Design of Embedded Systems," SystemC Technical Papers Collection, http://www.systemC.org

[16] M. AbdElSalam, A. Salem, G. Aly, "RTOS Modeling Using SystemC," in Proc. International Workshop for System On Chip (IWSOC) Conference, Canada, June 2002.

4

PART IV: SPECIFICATION FORMALISMS FOR PROVEN DESIGN

Chapter 22

AN ABSTRACT MODELING APPROACH TOWARDS SYSTEM-LEVEL DESIGN-SPACE EXPLORATION

F.N. van Wijk[1], J.P.M. Voeten[1], A.J.W.M. ten Berg[2]
[1]*Eindhoven University of Technology, Information and Communication Systems Group, Dept. of Elec. Eng., P.O. Box 513, 5600 MB Eindhoven, The Netherlands*
[2]*Philips Research Laboratories, Prof. Holstlaan 4, 5656 AA Eindhoven, The Netherlands*

Abstract: Integration of increasingly complex systems on a chip augments the need of system-level methods for specification and design. In the earliest phases of the design process important design decisions can be taken on the basis of a fast exploration of the design space. This paper describes an abstract modeling approach towards system-level design-space exploration, which is formal and flexible. It uses a uniform system model that contains both functional and architectural information. Disjunct, parameterizable resources represent the real-time behavior of the target architecture. Due to the expressiveness of the modeling language (POOSL), control as well as data oriented behavior can be specified in the functional part of the system model. Well-founded design decisions can be taken as a result of performance estimations that are based on Markov theory.

Key words: architecture exploration, design-space exploration, performance modeling, Parallell Object-Oriented Specification Language (POOSL), system-level design

1.   INTRODUCTION

In a traditional design trajectory, back-of-the-envelope calculations are directly followed by hardware and software development. It takes several iterations in the design process to reach an implementation that is functionally correct and satisfies the performance requirements. Those iterations consume large amounts of costly development time, especially since they occur in a phase of the design trajectory where there is already a

E. Villar and J. Mermet (eds.), System Specification and Design Languages, 267–282.
© 2003 *Kluwer Academic Publishers.*

lot of implementation detail involved. Furthermore, design decisions in this stage have only limited impact on the final implementation, as they are overshadowed by major design decisions taken earlier.

The design space is considered to be the collection of design alternatives, both feasible and infeasible, from which an 'optimal' alternative should get chosen. To speed-up the design process and to enable coverage of a larger part of the design space, it is necessary to take well-founded design decisions in an early phase of the design trajectory. System-level specification and design methods define frameworks for developing an executable model of the system, which describes the intended functionality in this early phase. The executable model allows the evaluation of system requirements against alternative conceptual solutions before starting the implementation of the system. System-level methods and tools should support fast exploration of different design alternatives with respect to hardware-software partitioning, selection of hardware components, choice of the communication resources, etcetera. A fast exploration can be attained by using abstract executable models.

Present-day signal processing systems are becoming more and more heterogeneous since they are often multi-functional (control oriented, programmable, software) on the one hand, and contain computation-intensive real-time calculations (data oriented, dedicated hardware) on the other hand. Our approach towards system-level design-space exploration is inspired by the Y-chart scheme for heterogeneous system design [1]. The most important aspect of the Y-chart is the separation of a functional specification from an architectural specification (on which the functionality is mapped), which gives one the flexibility to easily explore different design alternatives.

To construct an appropriate executable model, system-level methods for specification and design-space exploration should be based on well-defined modeling languages. In our approach we use the formal (i.e. mathematically defined) Parallel Object-Oriented Specification Language (POOSL) [2][3]. POOSL is a system-level modeling language for complex real-time hardware/software systems. The language is well suited for the construction of abstract executable models. Together with the application of the Y-chart scheme, this leads to an approach that enables a fast exploration of the design space at system level.

To introduce the concepts of our approach we use the example of a video filter application depicted in Figure 1. The Producer generates video frames, the Filter filters those frames on a line-by-line basis, and the Consumer reads in the filtered frames. They communicate via FIFO buffered channels, and their behavior is mapped onto Processing Resources P1 through P3. A possible design decision that has to be made is the choice between *alternative 1* and *alternative 2*. Is it necessary to separately map the Consumer onto resource P3, or can resource P1 be shared by the Producer and Consumer, saving us resource P3? And how does this decision affect the frame rate? We use our approach to find an adequate answer to these questions in this paper.

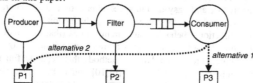

Figure 1. The Producer-Filter-Consumer (PFC) example

The remainder of this paper is organized as follows. In the next section we discuss related work. Section 3 gives a short overview of POOSL. The concepts of our approach towards system-level design-space exploration are described in Section 4. Simulation and performance analysis are the subjects of Sections 5 and 6. In Section 7 we present the results of our example design-space exploration exercise. Finally, in Section 8 we draw some conclusions and discuss future work.

2. RELATED WORK

Today, in the context of complex real-time heterogeneous system design, much effort is put in co-simulation. Typically, two rather low-level simulators are combined, one for simulating the programmable components that run the software and one for the dedicated hardware. Examples of co-simulators that run the hardware and software simulations separately are COSIM [4] and SYMPHONY [5], whereas e.g. POSEIDON [6] integrates hardware and software simulation. A serious drawback of these approaches is their inflexibility. The development of hardware blocks is totally different from the development of software tasks, which implies that it has to be decided in advance what parts should be implemented in hardware and what parts in software. This rules out the possibility to easily explore different hardware-software partitionings.

Apart from co-simulation, several groups are working on system-level architecture modeling and design-space exploration methodologies, e.g. the RASSP project [7] and SystemC [8]. A disadvantage of these approaches is that they are not really suited for exploration of the design space at a system's level of abstraction. To explore alternative architectures, serious rewriting of the system model is often needed, since the functionality is completely incorporated in the architecture model. At the highest level of abstraction however, RASSP does support performance modeling (only the timing behavior, not the functional behavior), but then the possibility to model data-dependent behavior is ruled out.

To support fast and easy design-space exploration, the Y-chart scheme has been developed [1]. In this scheme, a distinction is made between applications (functional behavior) and architectures (that eventually should execute this functional behavior). This allows easy exploration of different application-architecture combinations. Methods that separate functionality from architecture are e.g. CHINOOK [9][10] and COSY [11]. They both focus on the design of embedded systems composed of Intellectual Property (IP) components, such as microprocessors and programmable logic blocks. Characteristic of IP-based approaches is that they already go into quite some detail, both in their application models and IP-blocks, which makes performance analysis costly (in development and analysis time) and therefore limits the number of design alternatives that can be explored.

There are other approaches that also utilize the Y-chart scheme, but which use more abstract models to accomplish faster explorations. Examples are the work of Kienhuis [12] and the POLIS framework [13]. The Kienhuis approach is limited to a specific kind of dataflow architectures, and POLIS is best suited for reactive systems (control oriented). Combined data and control oriented behavior is not supported.

The work presented in this paper has mainly been inspired by the SPADE methodology [14]. SPADE is part of the ARTEMIS project that aims at the development of methods and techniques to support the design of highly programmable embedded media systems. SPADE also shows much resemblance with the architecture workbench of the MERMAID project [15]. Applications are modeled in YAPI (written in C/C++) [16], using Kahn Process Networks [17]. For the architecture specification generic building blocks are used, which are modeled in the Philips in-house cycle-based simulation system TSS. Application-architecture-mapping combinations are evaluated by means of a trace-driven simulation [18]. This implies that SPADE cannot be used to model control oriented behavior that contains non-determinism [19], while in abstract modeling the ability to model non-deterministic behavior is essential.

Our approach also complies with the Y-chart scheme that, in combination with the use of abstract models, allows us to do a fast exploration. In contrast with some of the above-mentioned approaches, in our approach both data and control-oriented behavior can be specified. Moreover, we use a single, formal model of computation to unambiguously specify the functional model, the architecture model, and the interfaces between them. Furthermore, the semantical model of Markov chains allows us to determine the accuracy of the performance metrics we obtain through simulation (see Section 5).

3. THE POOSL LANGUAGE

POOSL (Parallel Object-Oriented Specification Language) [2][3] is an expressive system-level modeling language in the sense that it allows us to represent a system in a succinct (i.e. compact) way. More important, POOSL enables a precise representation of the system, because it has a mathematically defined semantics. POOSL consists of a process part and a data part. The process part (processes and clusters) is founded on a real-time extension [3] of the process algebra CCS [20], whereas the data part is based upon the concepts of traditional sequential object-oriented programming languages such as Smalltalk and C++. The semantics is based on a two-phase execution model [21]; the state of a system can either change by asynchronously executing atomic (communication or data processing) actions (taking no time) or by letting the time pass (synchronously). The language is suitable for the verification of correctness properties and the evaluation of performance properties [22].

4. MODELING CONCEPTS

Figure 2 shows an outline of our approach towards system-level design-space exploration based on the Y-chart scheme [1]. It consists of a sequence of steps, which have to be taken in order to obtain a functionally correct system model that includes information about the target architecture:

1. Specify the Functional Model as a collection of tasks that communicate via unbounded FIFO channels.
2. Specify the Resource Model as a collection of (parameterizable) resources that are considered to be capable of executing the desired functional behavior.
3. Map the different entities of the Functional Model (i.e. tasks and channels including FIFO buffers) onto resource entities in the Resource Model.
4. Perform a quantitative analysis of the System Model as a whole, i.e. the Functional Model with its entities mapped onto resources in the Resource

Model. This simulation-based analysis results in performance figures of two kinds, those related to the Functional Model (e.g. throughput) and those related to the resources (e.g. utilization).

5. Iteratively explore the design space through modification of the Resource Model, the Mapping, or even the Functional Model on the basis of the performance figures (see the dashed lines in Figure 2) and repetition of step 4.

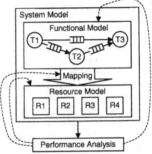

Figure 2. Modeling framework for system-level design-space exploration

In the remainder of this section, we elaborate on steps 1, 2, 3, and 5 by discussing one of the channels of the example introduced in Section 1 in more detail (see Figure 3). Step 4 will be detailed in Section 6.

4.1   Functional model

In the Functional Model (upper part of Figure 3) behavior is specified inside tasks (i.e. POOSL processes) that communicate asynchronously via conceptually unbounded FIFO channels. The task behavior can be specified using POOSL process and data statements. Through these statements both data and control-oriented behavior can be specified, including non-deterministic behavior (e.g. a possibly non-deterministic choice between available input messages). Therefore, our underlying model of computation is more expressive than e.g. Kahn process networks, which are deterministic and hence are not suited for systems containing both control and data.

The part of the Functional Model shown in Figure 3 consists of only two tasks (Producer and Filter) that are connected by a FIFO-buffered channel. The unbounded FIFO is split up in a number of unbounded FIFOs in concatenation. Conceptually, this concatenation of FIFOs exhibits the same behavior as one unbounded FIFO. The reason for this division is to enable a single channel in the Functional Model to be mapped onto a segmented channel that consists of multiple resources in the Resource Model (e.g. S1, C1, and S2).

The Functional Model supports automatic format conversion over FIFO-buffered channels, i.e. writing to and reading from a channel is allowed to take place with different data granularities (e.g. video frames vs. video lines). This format conversion takes place inside FIFO buffers (that are modeled as POOSL processes).

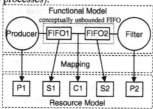

Figure 3. System model structure in POOSL

4.2   Resource model

The Resource Model of our approach differs significantly from existing approaches towards system-level architecture modeling (see e.g. [11], [12], [14] and [15]). Instead of an architecture model that consists of a number of architectural blocks that are mutually connected, our Resource Model consists of parameterizable resources that are not connected to each other at all. However, the way resources are connected is implicitly defined by the structure of the Functional Model (i.e. how tasks are mutually connected) in combination with the applied Mapping. Model time elapses in the resources, whereas tasks are synchronized with real-time through the Mapping (see Section 4.3).

Resource Models are built from three basic, parameterizable resource blocks:

- *Processing Resources (P).* The P-resource uses a task-level instruction table to assign resource-specific processing delays to (sub) tasks (e.g. FFT and DCT computations) that occur in the Functional Model.
- *Communication Resources (C).* The C-resource assigns a resource-specific communication delay to communications that take place over channels or channel segments in the Functional Model. The communication delays are proportional to the amount of information that is involved in a communication action. Therefore, the throughput of the

channel is implicitly defined by the delay of an elementary data unit (e.g. token or byte). This basic model can easily be extended to support characterization of communication through independent latency and throughput parameters.

- *Storage Resources (S)*. The capacity of the unbounded FIFOs in the Functional Model is restricted by the Storage Resources. Resource-specific storage capacity as well as access times and bandwidth can be modeled within Storage Resources.

The reason for using only three basic resources is that any hardware component (e.g. processor, bus, memory) can be thought to exist of a combination of those basic resources. In the example of Figure 3, the S1, C1 and S2 resources together may represent a buffered bus. More complex blocks can be obtained by combining several basic resources in POOSL clusters.

To ensure fairness in resource sharing of P-, C-, and S-resources, schedulers and arbiters are used. They are modeled in a separate layer on top of the Resource Model to allow easy exploration of different scheduling policies. Resources are shared by simply mapping multiple entities (i.e. tasks, FIFOs or channel segments) onto a single resource scheduler or arbiter.

4.3   Mapping

In other approaches based on the Y-chart scheme, mapping of a buffered communication channel is implicit, whereas the connections between architectural blocks are explicit. This means that only channel ports in the application model are mapped onto processor ports in the architecture model, thereby indirectly determining the channel in the architecture. In our approach the opposite holds: mapping of a buffered communication channel is explicit, whereas the connection of resources is implicit. This means that all segments of a channel in the Functional Model need to be mapped onto resources in the Resource Model, thereby indirectly modeling the structure of the resources (i.e. how they are mutually connected). This approach exhibits more flexibility, since resources can be combined without any restrictions to form any architecture, which facilitates easy exploration of different combinations of resources. Another advantage is that there is no separate flow of information through the architecture as in many other approaches.

Since we would like to support data as well as control oriented behavior, we have chosen for a direct mapping using request-response protocols, which leads to a strict coupling between the Functional Model and the

Resource Model. With a buffered mapping (i.e. used in trace-driven simulation techniques) it is impossible to accurately simulate non-deterministic behavior (that often is involved with control oriented behavior), whereas it can be simulated accurately if a direct mapping is used [19].

The mapping of the Functional Model onto the Resource Model is specified in terms of POOSL communication primitives (synchronous message passing over logical channels). Different mapping protocols are used for the three basic resources:

- *P*-resources receive task-level instructions from tasks that are mapped upon them. They subsequently let the time pass according to their task-level instruction tables, and then send back an acknowledgement. As an example, the essential P-resource behavior is depicted in Figure 4. In the context of the example introduced in Section 1, for the Filter task the instruction would be some filtering operation applied to a video line.
- *C*-resources receive requests for bandwidth from channel segments (in fact from the processes at the receiving ends) that are mapped upon them. They subsequently let the time pass proportional to the amount of data being transmitted, before an acknowledgement is sent back.
- *S*-resources typically use a two-phase mapping protocol. Before (a part of) the storage capacity of an S-resource can be used by a FIFO, it first has to be allocated to that FIFO. As soon as a FIFO does not need the allocated storage capacity anymore, it has to release it. Thus, S-resources exhibit blocking behavior.

```
Operate()()
| instruction: String; instructionDelay: Real |

task?resourceRequest(instruction);           // receive request
instructionDelay := instructionTable         // look-up delay in task-
    getDelay(instruction);                   // level instruction table
delay(instructionDelay);                     // wait
task!acknowledge;                            // send acknowledgement
Operate()().                                 // tail recursion
```

Figure 4. Essential P-resource behavior

5.   SIMULATION

Since exhaustive analysis is not feasible because of the state-space explosion problem, we use a simulation-based approach. As a consequence, the obtained performance figures are an estimation of their real value. In Section 6 we show how the accuracy of this estimation can be determined.

The SHESim tool [23] is used for modeling, validation and simulation of System Models that consist of a Functional Model, a Resource Model, and a Mapping. The System Model is simulated as a whole and, accordingly, the interfaces between the Functional Model and the Resource Model reside transparently within a single modeling environment. Because of the formal semantics of POOSL, the interface behavior is specified unambiguously. Apart from that, since we use abstract models, the approach is scalable towards larger target systems.

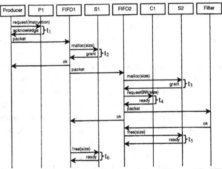

Figure 5. Message Sequence Chart resulting from simulation of the model of Figure 3

Simulation of the example in Figure 3 leads to a Message Sequence Chart (MSC) as shown in Figure 5. It demonstrates how the mapping protocols introduced in Section 4.3 result in an execution run. The MSC starts with a request of the Producer task to P-resource P1. P1 returns an acknowledgement after the time corresponding to the execution of the task-level instruction, e.g. the construction of a frame, has elapsed (t_1). The remainder of the MSC indicates how a packet (a chunk of information with a certain size) is transferred from the Producer to the Filter task. First, the packet is sent to FIFO1. FIFO1 can only store the packet after it has reserved storage capacity at S1 (malloc). FIFO1 sends back an acknowledgement (ok) to the Producer as soon as the packet is considered to be stored in S1 (grant, t_2). Only then the Producer may continue with its own behavior. After being stored in S1, the packet is immediately sent from FIFO1 to FIFO2, which has to reserve storage capacity at S2 before it can store the packet (t_3). When there is room available in S2, communication resource C1 is consulted (requestBW) to provide the required bandwidth on the transmission medium. C1 returns an acknowledgement (ready) as soon as the time needed for the transmission of the packet has elapsed (t_4). Only then the packet is actually stored in FIFO2 and an acknowledgement is sent back to FIFO1, so that FIFO1 can free its allocated memory (t_6). Also, as soon as the packet is stored in FIFO2, it is sent to the Filter. When the Filter has received the packet, it returns an acknowledgement to FIFO2, so that FIFO2 can free its allocated memory (t_5).

6. PERFORMANCE ANALYSIS

The semantics of POOSL is formally defined as a timed probabilistic labeled transition system [24] consisting of states and labeled transitions between states. In each state either a number of actions can occur, or the time can elapse for a certain amount of time, after which a new state is entered. If more than one action can be performed, a choice is made non-deterministically. If a choice is made and the corresponding action is performed, a probabilistic choice determines the next state. Time transitions are always deterministic and are taken with probability one. The semantical model obtained in this way can be considered to be a Markov decision process [22].

For performance analysis the Markov decision process is equipped with a reward structure. A reward is a real-valued function on the state-space of the model and the idea is that each time a certain state is visited a 'reward' as specified by the reward function is obtained. Many interesting performance metrics can be expressed in terms of long-run average rewards. Examples are utilization of processors, the average occupancy levels of buffers or the long-run variance in rate of a video stream.

To determine long-run average rewards, the Markov decision process is transformed into a Markov chain. This is done by a scheduler (also called policy), which resolves the non-determinism. A convenient scheduling policy is obtained by resolving non-determinism in a uniform way. Let $\{X_i \mid i \geq 1\}$ be the resulting Markov chain with state-space S and let r be a reward function. Then the long-run average reward is given by the limiting behavior of random variable $\frac{1}{n}\sum_{i=1}^{n} r(X_i)$. If some conditions of ergodicity are satisfied this variable converges strongly to a constant μ and this constant can be computed by $\mu = \sum_{s \in S} r(s)\pi_s$, where π_s denotes the equilibrium probability of state s.

In most practical cases, the state space of the Markov chain is far too large to be able to compute μ. In those cases a realization $\hat{\mu}$ of $\frac{1}{n}\sum_{i=1}^{n} r(X_i)$ can be used as a point estimation for μ and can be determined by simulation of the Markov chain. Further, the central limit theorem for Markov chains enables the determination of the accuracy of the estimation. The central limit theorem states that $\frac{\sqrt{n}}{\sigma}(\frac{1}{n}\sum_{i=1}^{n} r(X_i) - \mu)$ is asymptotically standard Normal. The central limit theorem allows us to derive confidence intervals that contain μ with a certain level of confidence (e.g. 95%). A confidence interval also lets us estimate an upper bound of the relative error $\frac{|\mu - \hat{\mu}|}{\mu}$. This is extremely important, since it gives us a criterion to end a simulation automatically when the relative error is smaller than the desired bound.

In case the non-determinism is not resolved, performance metrics do not give rise to a single performance figure but to a collection of figures, each figure corresponding to a different scheduling policy. Techniques exist to compute the lower and upper bounds of the collection of performance figures in case the number of states is not too large. Estimating the upper and lower bounds by simulation seems to be non-trivial however, and this topic will be subject of future research. This topic is important since it enables the analysis of models that deliberately abstract from lower-level design aspects.

7.   DESIGN-SPACE EXPLORATION EXERCISE

In this section we present the performance evaluation results of the design alternatives introduced in Section 1. A screenshot of the System Model in the SHESim tool is depicted in Figure 6. Besides the tasks and resources that have already been introduced, a task scheduler and a communication arbiter are situated in a separate layer. The tasks communicate over a shared C-resource C1. The design alternatives 1 and 2 are indicated by the thick lines in Figure 6.

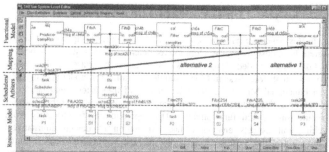

Figure 6. The PFC-example in the SHESim tool

For both alternatives we simulate the System Model and evaluate the long-run average utilization of the P-resources and the long-run average frame rate (i.e. throughput) at the Consumer. The results are presented in Table 1 and Table 2 respectively. Next to the point estimations, also the 90% confidence intervals are shown. Stop criterion of the simulations was a relative error bound of 1%.

Table 22-1. Performance evaluation results alternative 1 (90%) confidence

Metric	Point Estimation	Confidence Interval
Utilization P1	0.2253	[0.2232 , 0.2275]
Utilization P2	0.2246	[0.2224 , 0.2268]
Utilization P3	0.2249	[0.2227 , 0.2271]
Throughput Consumer (fps)	22.58	[22.38 , 22.78]

Table 22-2. Performance evaluation results alternative 2 (90% confidence)

Metric	Point Estimation	Confidence Interval
Utilization P1	0.4307	[0.4264 , 0.4349]
Utilization P2	0.2148	[0.2127 , 0.2169]
Throughput Consumer (fps)	21.33	[21.13 , 21.54]

From these results it can be concluded that the sharing of resource P1 leads to a small decrease of the throughput. The answer to the first question posed in Section 1 depends however on the minimum throughput requirement of the Functional Model. If the throughput of alternative 2 still meets this requirement, resource P3 can safely be omitted.

The design decision that has been discussed in this paper is a typical example of the kind of decisions that has to be taken during exploration of the design space, although for larger models taking design decisions gets more complicated because of the growing number of interdependencies.

8.   CONCLUSION AND FUTURE WORK

In this paper we have presented an abstract modeling approach towards system-level design-space exploration for heterogeneous systems. To be able to take well-founded design decisions in an early phase of the design process, an abstract System Model should be built. This model is comprised of a Functional Model (i.e. the functionality of the system) that has to be mapped onto a Resource Model (i.e. parameterizable resources that together execute this system functionality). The separation of functionality from architectural aspects enables a fast exploration of design alternatives. Simulation of the System Model yields performance metrics (e.g. resource utilization, throughput). Different design alternatives can be explored by means of changes to the Resource Model (add/remove resources, change parameter values) or the mapping on the basis of the results of the performance evaluation. We have illustrated our approach with a small example of a design-space exploration exercise.

Our approach offers the following combined advantages over existing approaches. In the Functional Model, both data and control-oriented behavior can be modeled. To support the modeling of non-determinism, which is an important issue in abstract modeling, a strict coupling between the Functional Model and the Resource Model has been chosen. This strict coupling made it possible to model resources as mutually disconnected blocks, which leads to more flexibility. There is no separate flow of information through the Resource Model. Furthermore, resource sharing is being covered in a natural way. On the basis of the formal semantics of the modeling language POOSL, the accuracy of the performance analysis results can be estimated. Finally, collections of scheduling policies may be evaluated without having to assume any deterministic scheduler.

The approach presented in this paper is the result of ongoing research. Extension of the approach to support stream-based functional models is an important issue, especially to accomplish more efficient data oriented modeling and performance analysis. Possibilities to extend the Resource Model with support for multiple resource scheduling policies need to be investigated. Furthermore, not all important performance metrics have been identified yet, as well as the way to represent them and analyze their accuracy. Currently the approach is being tested against a relevant industrial case.

REFERENCES

[1] B. Kienhuis, E. Deprettere, K. Vissers, and P. van der Wolf. "An Approach for Quantitative Analysis of Application-Specific Dataflow Architectures". In: L. Thiele, J. Fortes, K. Vissers, V. Taylor, T. Noll, and J. Teich, Eds., *Proc. of ASAP '97*, pp. 338-349. Los Alamitos, CA (U.S.A.): IEEE, 1997.

[2] P.H.A. van der Putten and J.P.M. Voeten. *Specification of Reactive Hardware/Software Systems*. Ph.D. thesis. Eindhoven (Netherlands): Eindhoven University of Technology, 1997.

[3] M.C.W. Geilen. "Real-time Concepts for a Formal Specification Language for Software/Hardware Systems". In: J.P. Veen, Ed., *Proc. of ProRISC '97*, pp. 185-192. Utrecht (Netherlands): STW, Technology Foundation, 1997.

[4] H. Hübert. *A Survey of HW/SW Cosimulation Techniques and Tools*. M.S. thesis. Stockholm (Sweden): Royal Institute of Technology, 1998.

[5] A.R.W. Todesco and T.H.-Y. Meng. "SYMPHONY: A Simulation Backplane for Parallel Mixed-Mode Co-Simulation of VLSI Systems". In: *Proc. of DAC '96*, pp. 149-154. New York, NY (U.S.A.): ACM, 1996.

[6] R.K. Gupta, C.N. Coelho Jr., and G. De Micheli. "Synthesis and Simulation of Digital Systems Containing Interacting Hardware and Software Components". In: *Proc. of DAC '92*, pp. 225-230. Los Alamitos, CA (U.S.A.): IEEE, 1992.

[7] C. Hein, J. Pridgen, and W. Kline. "RASSP Virtual Prototyping of DSP Systems". In: *Proc. of DAC '97*, pp. 492-497. New York, NY (U.S.A.): ACM, 1997.

[8] P.R. Panda. "SystemC – A Modeling Platform Supporting Multiple Design Abstractions". In: *Proc. of ISSS '01*, pp. 75-80. New York, NY (U.S.A.): ACM, 2001.

[9] P.H. Chou, R.B. Ortega, and G. Borriello. "The CHINOOK Hardware/Software Co-Synthesis System". In: *Proc. of the 8th Int. Symposium on System Synthesis*, pp. 22-27. Los Alamitos, CA (U.S.A.): IEEE, 1995.

[10] P. Chou, R. Ortega, K. Hines, K. Partridge, and G. Borriello. "IPCHINOOK: An Integrated IP-based Design Framework for Distributed Embedded Systems". In: *Proc. of DAC '99*, pp. 44-49. Piscataway, NJ (U.S.A.): IEEE, 1999.

[11] J.-Y. Brunel, A. Sangiovanni-Vincentelli, R. Kress, and W. Kruytzer. "COSY: A Methodology for System Design Based on Reusable Hardware and Software IP's". In: J.-Y. Roger, Ed., *Technologies for the Information Society*, IOS Press, pp. 709-716, 1998.

[12] A.C.J. Kienhuis. *Design Space Exploration of Stream-based Dataflow Architectures*. Ph.D. thesis. Delft (Netherlands): Delft University of Technology, 1999.

[13] F. Balarin, E. Sentovich, M. Chiodo, P. Giusto, H. Hsieh, B. Tabbara, A. Jurecska, L. Lavagno, C. Passerone, K. Suzuki, and A. Sangiovanni-Vincentelli. *Hardware-Software Co-Design of Embedded Systems - The POLIS Approach*. Dordrecht (Netherlands): Kluwer Academic Publishers, 1997.

[14] P. Lieverse, P. van der Wolf, K. Vissers, and E. Deprettere. "A Methodology for Architecture Exploration of Heterogeneous Signal Processing Systems". *Journal of VLSI Signal Processing Systems for Signal, Image, and Video Technology*, vol. 29, no. 3, pp. 197-207, 2001.

[15] A.D. Pimentel and L.O. Hertzberger. "An Architecture Workbench for Multicomputers". In: *Proc. of the 11th Int. Parallel Processing Symposium (IPPS '97)*, pp. 94-99. Los Alamitos, CA (U.S.A.): IEEE, 1997.

[16] E.A. de Kock, G. Essink, W.J.M. Smits, P. van der Wolf, J.-Y. Brunel, W.M. Kruijtzer, P. Lieverse, and K.A. Vissers. "YAPI: Application Modeling for Signal Processing Systems". In: *Proc. of DAC 2000*, pp. 402-405. New York, NY (U.S.A.): ACM, 2000.

[17] G. Kahn. "The Semantics of a Simple Language for Parallel Programming". In: *Proc. of IFIP '74*. North-Holland Publishing Co., 1974.

[18] R.A. Uhlig and T.N. Mudge. "Trace-Driven Memory Simulation: A Survey". *ACM Computing Surveys*, vol. 29, no. 2, pp. 128-170, 1997.

[19] S.R. Goldschmidt and J.L. Hennessy. "The Accuracy of Trace-Driven Simulations of Multiprocessors". *Performance Evaluation Review*, vol. 21, no. 1, pp. 146-157, 1993.

[20] R. Milner. *Communication and Concurrency*. Englewood Cliffs, New Jersey (U.S.A.): Pretence-Hall, 1989.

[21] X. Nicollin and J. Sifakis. "An Overview and Synthesis on Timed Process Algebras". In: K. Larsen and A. Skou, Eds., *Proc. of CAV '91*, pp. 376-398. Berlin (Germany): Springer-Verlag, 1991.

[22] J.P.M. Voeten. "Temporal Rewards for Performance Evaluation". In: *Proc. of PAPM '00 (ICALP Workshops 2000)*, pp. 511-522. Waterloo, Ontario (Canada): Carleton Scientific, 2000.

[23] M.C.W. Geilen and J.P.M. Voeten. "Object-Oriented Modelling and Specification using SHE". In: R.C. Backhouse and J.C.M. Baeten, Eds., *Proc. of VFM '99*, pp. 16-24. Eindhoven (Netherlands): Eindhoven University of Technology, 1999.

[24] R. Segala. *Modeling and Verification of Randomized Distributed Real-Time Systems*. Ph.D. thesis. Cambridge, MA (U.S.A.): Massachusetts Institute of Technology, 1995.

Chapter 23

MODELING TECHNIQUES IN DESIGN-BY-REFINEMENT METHODOLOGIES

Jerry R. Burch[1], Roberto Passerone[1], Alberto L. Sangiovanni-Vincentelli[2]

[1]*Cadence Berkeley Labs, Berkeley, CA, 94704.*

[2]*Department of EECS, University of California at Berkeley, Berkeley, CA 94720*

Abstract: Embedded system design methodologies that are based on the effective use of multiple levels of abstraction hold promise for substantial productivity gains. Starting the design process at a high level of abstraction improves control over the design and facilitates verification and synthesis. In particular, if we use a rigorous approach to link the levels of abstraction, we can establish properties of lower levels from analysis at higher levels. This process goes by the name of "design by refinement". To maximize its benefit, design by refinement requires a formal semantic foundation that supports a wide range of levels of abstraction. We introduce such a semantic foundation and describe how it can integrate several models for reactive systems.

Key words: Abstraction, refinement, heterogeneity, semantics.

1.   INTRODUCTION

Currently deployed design methodologies for embedded systems are often based on *ad hoc* techniques that lack formal foundations and hence are likely to provide little if any guarantee of satisfying a set of given constraints and specifications without resorting to extensive simulation or tests on prototypes. In the face of growing complexity and tighter time-to-market, cost and safety constraints, this approach will have to yield to more rigorous methods. The objective of the Metropolis project [2] is to provide a *design methodology and the software infrastructure* for embedded systems design, from specification to implementation, using methodologies such as platform-based design, communication-based design and successive refinement. The focus is on formal analysis and synthesis for *heterogeneous* systems, i.e.

E. Villar and J. Mermet (eds.), System Specification and Design Languages, 283–292.
© 2003 *Kluwer Academic Publishers.*

systems that use different modeling techniques for different parts of designs that interact with the real world. Metropolis is thus centered on its meta-model of computation, a set of primitives that are used to construct several different models. In this paper we lay the foundations for providing a denotational semantics for the meta-model. In particular we study the semantic domain of several models of computation of interest, and how relationships between these models can be established. To do so, we created a mathematical framework in which to express semantic domains in a form that is close to their natural formulation, i.e. the form that is most convenient for a model. The goal is to have a general framework that encompasses many models of computation, including different models of time and communication paradigms, and yet structured enough to give us results that apply regardless of the particular model in question. At the same time, the framework offers mathematical "tools" to help build new semantic domains from existing ones. Because the framework is based on algebraic structures, the results are independent of any particular design language, and therefore are not just specific to the Metropolis meta-model.

An important factor in the design of heterogeneous systems is the ability to flexibly use different levels of abstraction. Each part of the design undergoes changes in the level of abstraction during the design process and different abstractions are often employed for different parts of a design by way of different models of computation. Abstraction thus comes in different forms that include the model of computation, the scope (or visibility) of internal structure, or the model of the data. Thus, we provide a mathematical framework that allows the user to choose the best abstraction (semantic domain) for the particular task at hand. In this work, we concentrate on semantic domains for concurrent systems and on the relations and functions over those domains. We also emphasize the relationships that can be constructed between different semantic domains. This work is therefore independent of the specific syntaxes and semantic functions employed. Likewise, we concentrate on a formulation that is convenient for reasoning about the properties of the domain. As a result, we do not emphasize finite representations or executable models, which we defer for future work.

2.   RELATED WORK

In this section, we give a brief summary of the main approaches. We refer the reader to [6] for a more complete account of related work.

Several formal models have been proposed over the years [9] to capture one or more aspects of computation in embedded systems. Many models of computation can be encoded in the Tagged Signal Model [10]. In contrast, we describe a framework that is less restrictive in terms of what can be used to represent behaviors, and we concentrate on building relationships between the models that fit in the framework.

Our work shares the basic principles of the Ptolemy project [8] of providing flexible abstractions and an environment that supports a structured approach to heterogeneity. However, while in Ptolemy each model of computation is described operationally in terms of a common executable interface, we base our framework on a denotational representation and de-emphasize executability. Instead, we are more concerned with studying the process of abstraction and refinement in abstract terms. Process Spaces [11] are also an extremely general class of concurrency models. However, because of their generality, they do not provide much support for constructing new semantic domains or relationships between domains.

Our notion of conservative approximation is closely related to the Galois connection of an abstract interpretation [7]. In particular, the upper bound of a conservative approximation roughly corresponds to the abstraction function of a Galois connection. However, the lower bound of a conservative approximation appears to have no analog in the theory of abstract interpretations. Thus, conservative approximations allow non-trivial abstraction of both the implementation and the specification, while abstract interpretations only allow non-trivial abstraction of the implementation.

In the Rosetta language [1] domains of agents for different models of computation are described declaratively as a set of assertions in some higher order logic. In contrast, we are not concerned with the definition of a language, and we define the domain directly as a collection of elements of a set. In this sense, the approach taken by Rosetta seems more general. However, the restrictions that we impose on our models allow us to prove additional results that help create and compare the models.

3.   OVERVIEW

In the following sections we introduce our framework and concentrate on the basic principles underlying the definitions. We refer the reader to our previous publications [3][4][5][6] for an in depth presentation of specific examples of semantic domains and for a more formal presentation.

3.1   Traces and Trace Structures

The models of computation in use for embedded concurrent systems represent a design by a collection of agents (processes, actors, modules) that interact to perform a function. For any particular input to the system, the agents react with some particular execution, or behavior. In our framework we maintain a clear distinction between models of agents and models of individual executions. In different models of computation, individual executions can be modeled by very different kinds of mathematical objects. We always call these objects *traces*. A model of an agent, which we call a *trace structure*, consists primarily of a set of traces. This is analogous to verification methods based on language containment, where individual executions are modeled by strings and agents are modeled by sets of strings. However, our notion of trace is quite general and so is not limited to strings.

Traces often refer to the externally visible features of agents: their actions, signals, state variables, etc. We do not distinguish among the different types, and we refer to them collectively as a set of *signals W*. Each trace and each trace structure is then associated with an *alphabet $A \subseteq W$* of the signals it uses.

We make a distinction between two different kinds of behaviors: *complete* behaviors and *partial* behaviors. A complete behavior has no endpoint. A partial behavior has an endpoint; it can be a prefix of a complete behavior or of another partial behavior. *Complete traces* and *partial traces* are used to model complete and partial behaviors, respectively.

As an example we may consider traces that are suitable for modeling continuous time, synchronous discrete time and transformational (i.e. non-reactive) systems. In the first case, we might define a trace as a mapping that associates to each signal a function from continuous time (the positive reals) to an appropriate set of values (for example, the reals again). A partial trace in this case is a mapping that associates functions from only a closed time interval $[0, \delta]$, where $\delta > 0$. For an alphabet A, the complete traces are defined by $B_C(A) = A \rightarrow (\Re^+ \rightarrow \Re)$, while partial traces are defined by $B_p(A) = A \rightarrow ([0, \delta] \rightarrow \Re)$. We refer to these traces as *metric-time* traces.

An example of a trace suitable for synchronous discrete time systems is a sequence. In this example we assume the signals represent events that occur at distinct instants in time. The corresponding traces are sequences whose elements are subsets of the set of signals (events) available from the alphabet, i.e. $B(A) = (2^A)^\infty$ where the notation ∞ denotes both finite and infinite sequences. Here the partial traces are the finite sequences, while the complete traces are the infinite sequences. We refer to these traces as *synchronous* traces.

Unlike the previous two examples, a transformational system is only concerned with the initial and final state of a computation. If the signals are interpreted as state variables, a corresponding trace may be defined as a pair of mappings associating the state to its initial and final value (from a set of values V), respectively. The case of non-termination is modeled by adding a distinctive value $\perp$ to the set V. If we denote with $V_\perp = V \cup \{\perp\}$, partial and complete traces can be defined as $B(A) = (A \rightarrow V) \times (A \rightarrow V_\perp)$. We refer to these traces as *pre-post* traces.

Note that a given object can be both a complete trace and a partial trace; what is being represented in a given case is determined from context. For example, a terminating trace above can represent both a complete behavior that terminates or it can represent a partial behavior.

3.2   Trace Algebra and Trace Structure Algebra

In our framework, the first step in defining a model of computation is to construct a trace algebra. The carrier of a trace algebra contains the universe of partial and complete traces for the model of computation. The algebra also includes three operations on traces: *projection, renaming* and *concatenation*. These operations are defined to support common tasks used in design, like that of scoping, instantiation and composition of agents. The second step is to construct a trace structure algebra. Here each element of the algebra is a trace structure. Given a trace algebra a trace structure algebra is constructed in a fixed way. Thus, constructing a trace algebra is the creative part of defining a model of computation. A trace structure algebra includes four operations on agents: *projection, renaming, parallel composition* and *sequential composition.*

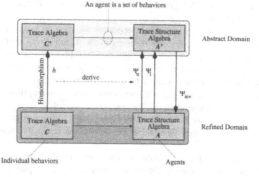

Figure 1. Algebras and their relationships

The relationships between trace algebras and trace structure algebras are depicted in Figure 1. This figure also shows the relationships between different algebras that we will discuss later in the paper.

The first operation of trace algebra is called *projection*, and consists in retaining from a trace only the information related to certain signals. The projection operation *proj* takes as argument the set of signals B that should be retained. For metric-time traces, projection corresponds to restricting the mapping from the set A of signals to a subset B. This operation is defined similarly for pre-post traces. Conversely, for synchronous traces projections consists of removing the events not in B from each element of the sequence. Projection on trace structures (agents) corresponds to hiding the internal signals, and can therefore be seen as the natural extension to sets of the corresponding operation on individual traces. In other words, the scope of the hidden signals is limited to the agent they belong to.

The second operation is called *renaming*, and consists in changing the names of the visible elements of a trace or an agent. The renaming operation *rename* takes as argument a renaming function r that maps the elements of the alphabet A into a new alphabet C. The function r is required to be a bijection in W to avoid conflicts of names and potentially a change in the behavior of the agents. In all examples, renaming corresponds to a substitution of signals, and is easily defined in terms of the renaming function r or of its inverse. As for projection, renaming of trace structures can be seen as the natural extension to sets of traces of the corresponding operation on individual traces. The effect is that of a renaming of all the signals in the agent: this process corresponds to that of *instantiation* of a master agent into its instances.

Projection and renaming, seen as operators for scoping and instantiation, are common operations that are meaningful to all models of computation. For all trace algebras and all trace structure algebras we require that the operations on traces and trace structures satisfy certain properties. These properties ensure that the operations behave as expected given their intuitive meaning. In addition, we can use these properties as assumptions to prove results that are independent of the particular model of computation in question. These results provide powerful tools that the designer of the model can use to prove general facts about the model and about its relationships with other models.

In the algebra of trace structures we introduce the additional operation of parallel composition. Note that this operation has no counterpart in the trace algebra. Intuitively, parallel composition corresponds to having several agents run concurrently by sharing some common signals. The result of the parallel composition is one agent that alone acts as the combination of the agents being composed. Let $T_1 = (A_1, P_1)$ and $T_2 = (A_2, P_2)$ be two agents that we want to compose, where A_1 and A_2 are the alphabets, and P_1 and P_2 the set of traces. The alphabet of the parallel composition $T = T_1 \parallel T_2$ must include all signals from T_1 and T_2, so that $A = A_1 \cup A_2$. The set P of traces of T must be "compatible" with the restrictions imposed by the agents being composed. Thus if x is a trace of T with alphabet A, then its projection $proj(A_1)(x)$ on the alphabet of T_1 must be in P_1, and the projection $proj(A_2)(x)$ on the alphabet of T_2 must be in P_2. The set of traces in T must be maximal with respect to that property. Formally:

$$P = \{x \in B(A) \mid proj(A_1)(x) \in P_1 \wedge proj(A_2)(x) \in P_2\}.$$

Similarly to projection and renaming, parallel composition of agents must satisfy certain properties. For example we require that it be commutative and associative. A fundamental result of this work is that the properties of the trace algebra are sufficient to ensure that the corresponding trace structure algebra satisfies its required properties.

While parallel composition is at the basis of concurrent models of computation, in other models the emphasis may be on a "sequential execution" of the agents. For these models we introduce a third operation on traces called *concatenation*. In the case of synchronous traces, concatenation corresponds to the usual concatenation on sequences. Similarly we can define concatenation for metric-time traces. For pre-post traces, concatenation is defined only when the final state of the first trace matches the initial state of the second trace. The resulting trace has the initial state of the first component and the final state of the second. Note that the information about the intermediate state is lost.

Similarly to the other operations, concatenation must also satisfy certain properties that ensure that its behavior is consistent with its intuitive interpretation. For example we require that it be associative (but not commutative!), and that it behaves consistently when used in combination with projection and renaming. Concatenation induces a corresponding operation on trace structures that we call *sequential composition* by naturally extending it to sets of traces. A more detailed account of sequential composition can be found in [5].

3.3 Refinement and Conservative Approximations

In verification and design-by-refinement methodologies a specification is a model of the design that embodies all the possible implementation options. Each implementation of a specification is said to *refine* the specification. In our framework, each trace structure algebra has a refinement order that is based on trace containment. We say that an agent T_1 refines an agent T_2, written $T_1 \subseteq T_2$, if the set of traces of T_1 is a subset of the set of traces of T_2. Intuitively, this means that the implementation T_1 can be substituted for the specification T_2. Proving that an implementation refines a specification is often a difficult task. Most techniques decompose the problem into smaller ones that are simpler to handle and that produce the desired result when combined. To make this approach feasible, the operations on the agents must be monotonic with respect to the refinement order. The definitions given in the previous section make sure that this is the case for our semantic domains.

An even more convenient approach to the above verification consists of translating the problem into a different, more abstract semantic domain, where checking for refinement of a specification is presumably more efficient. A *conservative approximation* is a mapping of agents from one trace structure algebra to another, more abstract, algebra that serves that purpose. The two trace structure algebras do not have to be based on the same trace algebra. Thus, conservative approximations are a bridge between different models of computation (see Figure 1).

A conservative approximation is actually composed of two mappings. The first mapping is an upper bound of the agent: the abstract agent represents all of the possible behaviors of the agent in the more detailed domain, plus possibly some more. This mapping is usually denoted by Ψ_u. The second is a lower bound: the abstract agent represents only possible behaviors of the more detailed one, but possibly not all. We denote it by Ψ_l.

Conservative approximations are abstractions that maintain a precise relationship between verification results in the two trace structure algebras. In particular, a conservative approximation is defined to preserve results related to trace containment, such that if T_1 and T_2 are trace structures, then $\Psi_u(T_1) \subseteq \Psi_l(T_2)$ implies that $T_1 \subseteq T_2$. When used in combination, the two mappings allow us to relate results in the abstract domain to results in the more detailed domain. The conservative approximation guarantees that this will not lead to a false positive result, although false negatives are possible.

Defining a conservative approximations and proving that it satisfies the definition can sometimes be difficult. However, a conservative approximation between trace structure algebras can be derived from a homomorphism between the underlying trace algebras.

A homomorphism h is a function between the domains of two trace algebras that commutes with projection, renaming and concatenation. Consider two trace algebras C and C'. Intuitively, if $h(x) = x'$ the trace x' is an abstraction of any trace y such that $h(y) = x'$. Thus, x' can be thought of as representing the set of all such y. Similarly, a set X' of traces in C' can be thought of as representing the largest set Y such that $h(Y) = X'$, where h is naturally extended to sets of traces. If $h(X) = X'$, then $X \subseteq Y$, so X' represents a kind of upper bound on the set X. Hence, if A and A' are trace structure algebras constructed from C and C' respectively, we use the function Ψ_u that maps an agent with traces P in A into the agent with traces

$h(P)$ in A' as the upper bound in a conservative approximation. A sufficient condition for a corresponding lower bound is: if $x \notin P$, then $h(x)$ is not in the set of possible traces of $\Psi_l(T)$. This leads to the definition of a function $\Psi_l(T)$ that maps P into the set $h(P) - h(B(A) - P)$. The conservative approximation $\Psi = (\Psi_l, \Psi_u)$ is an example of a *conservative approximation induced by h.* A slightly tighter lower bound is also possible (see [3]).

Thus, one need only construct two models of individual behaviors and a homomorphism between them to obtain two trace structure models along with a conservative approximation between the trace structure models.

3.4   Inverses of Conservative Approximations

Conservative approximations represent the process of abstracting a specification in a less detailed semantic domain. Inverses of conservative approximations represent the opposite process of refinement.

Let A and A' be two trace structure algebras, and let Ψ be a conservative approximation between A and A'. Normal notions of the inverse of a function are not adequate for our purpose, since Ψ is a pair of functions. We handle this by only considering the T in A for which $\Psi_u(T)$ and $\Psi_l(T)$ have the same value T'. Intuitively, T' represents T exactly in this case, hence we define $\Psi_{inv}(T') = T$. When $\Psi_u(T) \neq \Psi_l(T)$ then Ψ_{inv} is not defined.

The inverse of a conservative approximation can be used to embed a trace structure algebra at a higher level of abstraction into one at a lower level. Only the agents that can be represented exactly at the high level are in the image of the inverse of a conservative approximation. We use this as part of our approach for reasoning about heterogeneous systems that use models of computation at multiple levels of abstraction. Assume we want to compose two agents T_1' and T_2' that reside in two different trace structure algebras A_1 and A_2. To make sense of the composition, we first define a third, more detailed trace algebra that has homomorphisms into the other two. Thus we can construct a third, more detailed, trace structure algebra A with conservative approximations induced by the homomorphisms. The inverse of these conservative approximations are used to map T_1' and T_2' into their corresponding detailed models T_1 and T_2. The composition then takes place in the detailed trace structure algebra.

4.   CONCLUSIONS

The goal of trace algebras is to make it easy to define and to study the relationship between the semantic domains for a wide range of models of computation. All the models of importance "reside" in a unified framework so that their combination, re-partition and communication may be better understood and optimized. This unified approach will provide a designer a powerful mechanism to actually select the appropriate models of computation for the essential parts of his/her design.

Our representation of agents is denotational, in that no rule is given to derive the output from the input. The algebraic infrastructure allows us to formalize a semantic domain in a way that is close to a natural semantic domain for a model of computation. In addition it introduces additional concepts such as hierarchy, instantiation and scoping in a natural and consistent way. In particular we are concentrating on using the concept of a conservative approximation to study the problem of heterogeneous interaction.

REFERENCES

[1]	The Rosetta web site. http://www.sldl.org.
[2]	F. Balarin, L. Lavagno, C. Passerone, A. L. S. Vincentelli, M. Sgroi, and Y. Watanabe. Modeling and designing heterogeneous systems. In J. Cortadella and A. Yakovlev, editors, Advances in Concurrency and System Design. Springer-Verlag, 2002.
[3]	J. R. Burch. Trace Algebra for Automatic Verification of Real-Time Concurrent Systems. PhD thesis, School of Computer Science, Carnegie Mellon Univ., Aug. 1992.
[4]	J. R. Burch, R. Passerone, and A. Sangiovanni-Vincentelli. Overcoming heterophobia: Modeling concurrency in heterogeneous systems. In M. Koutny and A. Yakovlev, editors, Application of Concurrency to System Design, 2001.
[5]	J. R. Burch, R. Passerone, and A. L. Sangiovanni-Vincentelli. Using multiple levels of abstraction in embedded software design. In T. A. Henzinger and C. M. Kirsch, editors, 1st International Workshop, EMSOFT 2001, vol. 2211 of LNCS. Springer-Verlag, 2001.
[6]	J. R. Burch, R. Passerone, and A. L. Sangiovanni-Vincentelli. Modeling techniques in design-by-refinement methodologies. In Proceedings of the Sixth Biennial World Conference on Integrated Design and Process Technology, June 23-28 2002.
[7]	P. Cousot and R. Cousot. Abstract interpretation: a unified lattice model for static analysis of programs by construction or approximation of fixpoints. In Conference Record of the Fourth Annual ACM SIGPLAN-SIGACT Symp. on Principles of Programming Languages, pages 238--252, Los Angeles, California, 1977.
[8]	J. Davis II, et al. Heterogeneous concurrent modeling and design in java. Technical Memorandum UCB/ERL M01/12, EECS, Univ. of California, Berkeley, Mar. 2001.
[9]	S. Edwards, L. Lavagno, E. Lee, and A. Sangiovanni-Vincentelli. Design of embedded systems: Formal models, validation, and synthesis. Proc. of the IEEE, 85(3):366--390, Mar. 1997.
[10]	E. A. Lee and A. L. Sangiovanni-Vincentelli. A framework for comparing models of computation. IEEE Transactions on CAD, 17(12):1217--1229, Dec. 1998.
[11]	R. Negulescu. Process spaces. In C. Palamidessi, editor, CONCUR, volume 1877 of Lecture Notes in Computer Science. Springer-Verlag, 2000.

Chapter 24

DESIGN OF HIGHLY PARALLEL
ARCHITECTURES WITH ALPHA AND HANDEL

Florent de Dinechin, M. Manjunathaiah, Tanguy Risset and Mike Spivey
INRIA/LIP ENS-Lyon 69364 Lyon, France and OUCL, Parks Road, Oxford OX1 3QD, UK

Abstract: We propose a bridge between two important parallel programming paradigms: data parallelism and communicating sequential processes (CSP). Data parallel pipelined architectures obtained with the Alpha language can be embedded in a control intensive application expressed in CSP-based Handel formalism. The interface is formally defined from the semantics of the languages Alpha and Handel. This work will ease the design of compute intensive applications on **FPGAs.**

Key words: data parallelism, communicating sequential processes, FPGA, hardware compiling

1.   INTRODUCTION

Increasingly high-level approaches are being adopted in the design/codesign of high-performance embedded systems due to the complexity in implementing complete systems in silicon or as a mixture of hardware and software. Many applications are both control and compute intensive in nature and hence highly parallel solutions are sought to meet real-time constraints. Constructing a design, which maximizes the available parallelism, is a difficult task. Additionally, designers are faced with the problem of designing complex systems more rapidly. The natural solution to this problem is to use automatic or semi-automatic compilation techniques from high-level specifications. One of the main difficulties in this process is the definition of a correct computation model that can express high-level functional specifications as well as final target designs and implementations. Existing tools use a particular computation model that can express parallelism (because hardware is targeted for performance) for a specific

E. Villar and J. Mermet (eds.), System Specification and Design Languages, 293–302.
© 2003 *Kluwer Academic Publishers.*

granularity [1]. Among the existing models, two are especially important: the data parallel model and the communicating sequential processes model.

In this paper, we propose a way of easily and rapidly generating correct highly parallel designs by bridging the two models. The idea is to use the systolic synthesis methodology, which has been implemented in the MMAlpha tool, to provide correct specifications of pipelined architectures and to use the hardware compilation methodology provided by the Handel compilers to automatically translate them into hardware circuits. As these architectures are expressed in the Alpha language which has a very precise semantics, we are able to define a translation into Handel which can be guaranteed to be correct, provided that the design has some well defined characteristics. This integrated design methodology will improve the design time of implementing pipelined architectures on FPGAs.

2.   HARDWARE COMPILATION PARADIGMS

In this section we highlight the differences between concepts used in the data parallel model and the communicating sequential processes model (respectively illustrated by Alpha and Handel). We show that the design methodologies and the resulting hardware of these two tools are strongly influenced by their underlying computation models. We illustrate this by the Finite Impulse Response (FIR) filter, which is a simple convolution filter, used in many digital signal processing algorithms (correlation, adaptive filtering, etc.). The FIR algorithm is the following: given a possibly infinite sequence of values $x(n)$, $(n \geq 0)$ and N weights $w(i)$, $(0 \leq i \leq N\text{-}1)$, compute the sequence $y(n)$, $(n \geq N\text{-}1)$ where:

$$y(n) = \sum_{i=0}^{N-1} x(n\text{-}i)w(i) \quad . \quad (1)$$

2.1 Data parallel formalism

The data parallel model led to the definition of HPF and to the development of methodologies for the design of highly parallel or pipelined architecture [2]. The applications targeted, which are usually compute intensive, are composed of regular and repetitive operations on large sets of data (signal processing, multimedia, etc.). Research at Irisa (Rennes) produced the MMAlpha system [3] based on the Alpha language [4], which is dedicated to the hardware synthesis of regular (systolic-like) architectures from high level (functional) specifications. A fundamental feature of this methodology is that the parallelism is initially implicit and the design process provides transformations to introduce concurrency in a controlled and (semi)- automatic way. The Alpha specification of the FIR filter is shown in Figure 1. M is the number of inputs $x(n)$, and all variables are represented by 16-bit integers (as we will see later, we can also express the algorithm at the bit level). This specification is obtained very easily from the equation (1), it has no specified order of execution.

```
system fir:{N,M | 3<=N<=M-1}
  (x : {n | 1<=n<=M} of
                integer[S,16];
  w : {i | 0<=i<=N-1} of
                integer[S,16])
returns (res : {n | N<=n<=M}
                of integer[S,16]);
var
  Y : {n,i | N<=n<=M; -1<=i<=N-1}
                of integer[S,16];
let
  Y[n,i] = case
    { | i=-1} : 0[];
    { | 0<=i} : Y[n,i-1] +
                w[i] * x[n-i];
  esac;
  res[n] = Y[n,N-1];
tel;
```

```
void main (void) {
  int 16 x[6];
  int 16 w[3];
  int 16 Y[4,4];
  int 16 res[4];

  for (n = 0; n <= 5; ++n) {
    Y[n][0]=0;
    for (i = 1; i <= 2; ++i) {
      Y[n][i]=Y[n][i-1]+w[i]*x[n-i]
    }
  }
}
```

*Figure 1.*Specification of the FIR filter in Alpha (left, this specification is directly obtained by serializing the summation in equation (1)), and a possible Handel program for it (right, for values N=3 and M=6).

The MMAlpha system [3] can automatically generate a systolic architecture expressed in VHDL from this specification. This process involves several refinement steps in which the initial program is transformed into equivalent ones. The resulting architecture, sketched in Figure 2, is itself completely described (both its structure and its behaviour) *in the Alpha language*. One intermediate form of the program is the *space-time representation* of the program (hereafter referred as Alpha0 format). Here is an example of an Alpha equation in space-time representation:

```
Y[t,p] =   case
  { | p=-1} : 0[];
  { | 0<=p} : Y[t-1,p-1] + wPipe[t-1,p] * xPipe[t-1,p];
  esac ;
```

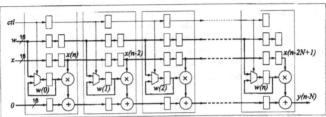

Figure 2. Register transfer level description semi-automatically derived from the Alpha specification of Figure 1 (Left) with the systolic design methodology implemented in the MMAlpha environment

In this representation, index t can be interpreted as time and index p can be interpreted as a processor number. It means that the first index of each variable represents a counter on a global clock regulating the array. This assumption leads to a natural hardware target: a globally synchronous digital circuit in which all the registers are controlled by a common virtual clock. The virtual clock can be implemented using the clock enable signal on Xilinx FPGA for instance. This strong assumption allows the designer to implement regular designs very efficiently, with minimal control overhead (control signal are usually pipelined in the array). Moreover, this control mechanism can be generated automatically because of restrictions on input format (static control). This whole refinement process can be formally proven.

2.2   Communicating sequential processes

The communicating sequential processes model [5] led to two important languages: Occam for concurrent software design [6] and Handel for concurrent hardware design [7]. The Handel family of languages q(HandelC [7], HandelB [8]) is particularly targeted at implementing programs in hardware. The Handel formalism is well suited for expressing complex control structures. In contrast with the data parallel model, the concurrency is explicit and the designer has to specify the concurrency using the language features. This often makes it difficult to design highly parallel systems as the complexity of the design increases a lot with the amount of parallelism and communication. Therefore for a class of concurrent designs, such as pipelined architectures, it would be interesting to combine the two design methodologies.

A possible Handel specification of the FIR of equation (1) is shown in the right of Figure 1 (we used the HandelC formalism here). This specification is very close to a C formulation but it does not contain any parallelism (Of course one could come up with an equivalent parallel program, but that is what we aim to achieve using the data parallel approach and we demonstrate it in a later section). The hardware compiled from the program of Figure 1 with the Handel compiler will respect some important guidelines [7,8]. The compiler generates registers as resources for (static) arrays and a RAM module (memory and addressing logic) for dynamic arrays. A key feature of the language is that each assignment will take exactly one clock cycle, *provided it can be executed*. In practice, it means that each variable will be stored in a register, which is itself controlled by a boolean signal. A global clock is continuously applied to all registers (synchronous design) but a *control token* mechanism is added to express the flow of control (control states) dynamically at run-time. The control tokens are used to schedule the assignment and communication statements in the language [7]. This control token acts like a clock enable.

3.   COMBINING THE MODELS

To illustrate the difference in the two models, we have shown on Figure 5 a simple piece of hardware as it would be implemented from the MMAlpha system (a) and from the Handel compiler (b). A 5 bit signal r gets the value of one of the two registers: a or b. Note that in the implementation from MMAlpha, there is a simple global clock enable, but the control of the multiplexer is done explicitly while in the Handel mechanism, the clock enable may be different for each signal and the control of the multiplexer is done using these clock enable signals. Note also that, if the MMAlpha hardware is slightly more difficult to generate (because it implies the correct setting of the ctrl signal), the resulting complexity is lower than the Handel implementation complexity (because the clock enable mechanism is simpler).

Another important difference between the two languages is the status of variables. In Alpha, all variables represent *signals* and any piece of hardware is represented by an operation between signals: operators (for combinatorial logic) or dependencies (for registers or simple connexion between signals). In Handel, all variables represent registers, however, Handel also permits to express signals with the always construct in HandelB and the signal type in HandelC.

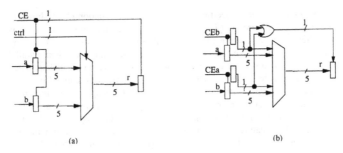

(a)   (b)

Figure 3. Different hardware generated from MMAlpha (a) and Handel (b) having the same behaviour (provided ctrl is correctly set up)

We would like to highlight the following: it is possible to define equivalence between hardware described in Alpha and hardware described in Handel. Although implementations will be different, the two designs will have the same behaviour. For instance, in Figure 5 both designs will have the same behaviour (provided the control signal ctrl is correctly set up in the MMAlpha hardware generation). Note that the equivalence between Handel hardware and MMAlpha hardware will be very difficult to prove if the Handel program uses channels, hence we will not use channels in the translation. We will use channels to synchronize Input and Output with the calling Handel Program.

If we want to translate an architecture expressed in Alpha0 into Handel we have to solve to two following problems: translate registers (which are operators in Alpha) into variables and minimize the control overhead that will be introduced by the local clock enable Handel mechanism. We will explain that in section 3.2.

In the following section, we describe how to bridge the Alpha and Handel languages in deriving pipelined designs. As it involves several tools in the design process, we refer to this as *co-design*, although it may not be hardware/software co-design but rather hardware/hardware co-design. As in a classical co-design framework, we first need a co-simulation step in which the functionality of the design is verified and we then need a co-synthesis step.

3.1   Co-simulation

The first architectural description is obtained after the *space-time* transformation. Co-simulation can be applied as soon as the Alpha program has been scheduled. Scheduling attempts to find an execution date for each computation (The scheduled equation defining Y is shown in section 2.1). Here, each Alpha equation takes one clock cycle hence the scheduled program can be easily translated into Handel, the main difficulty being to translate polyhedral domains into loops. This non-trivial problem is already solved in the Alpha to C compiler [9]). For this example, the resulting Handel program will contain an output sequential replicator (loop on index t) and many inner parallel replicators, (loops on index p).

We refer to this translation as "naive" because the efficiency of the resulting Handel program is not addressed, only a semantically equivalent program is derived. Indeed, this translation does not attempt to reuse memory. As Alpha is a single assignment language, the amount of memory used in the resulting Handel program can be quite large. However, it has the same functional behavior as the final architecture. Functional validity can be verified by simulating the translated program using different simulators (source level and circuit level simulators) and checking it against the execution of the C code output from the MMAlpha code generator.

3.2   Co-synthesis

After the functional validation of the design in both MMAlpha and Handel has been established, we look for a more efficient implementation of the Alpha architecture in the Handel formalism. More specifically, we aim to translate a more detailed design from the Alpha0 format in which control signals are made explicit. In an Alpha0 program, some equations express registers between two Alpha signal of the same processor (for instance, referring to the Alpha0 equation of section 2.1: `Y[t][p] = ...wPipe[t-1][p]..`) and some equations express registers between two Alpha signals of different processors (e.g. `Y[t][p]=...Y[t-1][p-1]..`). In the translation in Handel of this equation, this Y signal will give rise to a signal (called Y) and a register arrays indexed by p (called YReg[p]):

```
signal int 16 Y;
int 16 YReg[N], wPipeReg[N], xPipeReg[N];
...
while (True) {
 par (p=0;p<=N-1;p++)
```

```
par { Y = (p==0) ? 0 : ((p>0) ? YReg[p] : 0)
        YReg[p] = YReg[p-1]+wPipeReg[p]*xPipeReg[p];
    }
...
```

Note that we do not need an array of signals because we know (as MMAlpha derives systolic architectures) that any connexion between different processors will be associated to a register (there is no broadcast). hence, it is always possible to use signals inside a processor, and registers in the communications between processors. This fairly simple translation scheme can be applied to all Alpha0 equations. The resulting Handel program will be composed of a sequential while(True) enclosing the body of the program which should take one clock cycle. The body of the program itself is composed of a replicated par indexed by the processor index p (for more complicated examples there may be several processor indices) . The different behaviours of different processors is selected by a conditional selection. Finally the input to the systolic design is done with a Handel channel. The architecture has to receive two streams of data (x and w) together with a control signal generated by the MMAlpha design flow.

The architecture will be synthesized with Handel hardware control mechanism (one control latch for each register), but as the body of the program takes only one clock cycle, the Handel compiler can optimize the control and realize that it can just broadcast the clock enable signal to all the registers of the design. This parallel architecture can now easily be incorporated into a more complex design involving control dominant computations since the Handel language has sufficient expressive power for both styles of designs

4.   COMBINING MACRO AND MICRO PARALLELISM

The architecture, which has been used for illustrating our translation, is pipelined at the word level, which is relatively coarse grain. Notably, the implementation of the multipliers is undefined in the resulting abstract architecture. The rate at which the array can be pipelined will be determined by the slowest element of the array, and multiplication can pose a bottleneck. Hence, in order to achieve fast cycle times (computing at the rate of addition) a refinement of the multiplication operations is necessary.

A natural solution is to obtain a pipelined design for the multiplier and this can be realized as a program transformation ahead of hardware compilation. One of the advantages of our approach is to allow the generation of finer grained pipelines as high-level transformations in the design process, typically with a range of pipelined multipliers, without increasing the complexity of the design process. The idea here is to express the multiplication as an Alpha program at the boolean level and apply the same transformation steps as in the macro pipeline generation to derive micro pipelines. This idea is particularly relevant in the context of FPGA compilation, because most recent FPGA chips provide specific hardware for fast carry propagation. Therefore, the back-end tool has little choice when implementing additions, and can be expected to do it well. Conversely, the design space for multipliers is quite complex, and requires design choices involving low-level considerations. Our approach is to lift up these design choices to the specification level, and let MMAlpha handle the details of pipeline management.

We have derived a bit level specification of the FIR filter in Alpha (it basically consists in adding one dimension to all variables: the bits and changing operators into calls to library). The script used to derive the word level architecture can be re-used to derive the bit level architecture. The resulting architecture will contain only boolean operations. If the multiplier is completely pipelined, the resulting clock cycle will be not more than that one of an adder thus increasing the throughput of the array.

5. CONCLUSION

We have explained our methodology, which uses two important tools for obtaining highly parallel hardware designs. These hardware designs will also contain a complex control part, and this is the major originality of this work. There are many advantage of Handel over conventional hardware description languages for high-level design. Handel (as well as Alpha) has clean semantics allowing high confidence in correctness of translation between Alpha and Handel. Handel provides high level and low-level simulation which is very useful for debugging. Finally, the ability to express control-intensive part of application in high-level form is very important to provide correct designs.

In this paper, no experimental results are presented yet because this is currently on-going work, but we believe that the idea of collaboration between data parallel model and communicating sequential process model is an important subject that has to be discussed by the scientific community.

6.   REFERENCES

[1] E.A. Lee et al. Overview of the ptolemy project. Technical Report UCB/ERL No. M99/37, University of California, Berkeley, july 1999.

[2] Guy-René Perrin and Alain Darte, editors. *The Data Parallel Programming Model*, volume 1132 of *LNCS Tutorial*. Springer Verlag, 1996

[3] A.C. Guillou, P. Quinton, T. Risset, and D. Massicotte. High Level Design of Digital Filters in Mobile Communications. DATE Design Contest 2001, March 2001. Second place, available at http://www.irisa.fr/bibli/publi/pi/2001/1405/1405.html.

[4] Api-Cosi. MMAlpha Reference Manual, 1997.

[5] C. A. R. Hoare. *Communicating Sequential Processes*. Prentice-Hall, Englewood Cliffs, NJ, 1985.

[6] David May. Occam. *Computer Bulletin*, 1(1):14, March 1985.

[7] I. Page. Constructing hardware-software systems from a single description. *Journal of VLSI signal processing*, 12:87--107, 1996.

[8] J.M. Spivey. Deriving a Hardware Compiler from operational semantics. Technical Report available from http://oldwww.comlab.ox.ac.uk/oucl/users/mike.spivey/, Oxford University Computing Laboratory, 1997.

[9] F. Quilleré, S. Rajopadhye, and D. Wilde. Generation of Efficient Nested Loops from Polyhedra. *International Journal of Parallel Programming*, 28(5):469--498, 2000.

Chapter 25

MTG* AND GRAY BOX
Modeling dynamic multimedia applications with concurrency and non-determinism

Stefaan Himpe[1], Geert Deconinck[1], Francky Catthoor[2], and Jef van Meerbergen[3]
[1]Katholieke Universiteit Leuven, [2]IMEC, [3]Philips Natlab

Abstract: In this paper we introduce the MTG* model - a successor to the MultiThread Graph (MTG) model - that intends to expose task level parallelism and dynamic and non-deterministic behavior in advanced multimedia and telecom applications. After informal introduction of some core model concepts, we focus on the constructs in the MTG* model to represent dynamic and non-deterministic behavior in such applications, based on the new concept of an MTG* thread frame. Extensions of MTG* to arrive at a Gray-Box model are introduced as a way to manage large and complex designs with dynamic concurrent behavior, without loosing a global view on the system behavior. We briefly touch upon how the Gray-Box model enables modern design methodologies to take into account the dynamic characteristics, while trying to make a power consumption efficient implementation of the application on an embedded multi-processor hardware platform. Finally, a prototype Gray-Box simulator is demonstrated on a dynamic video recorder test vehicle.

key words: modeling, dynamism, concurrency, non-determinism, design methodology

1.   INTRODUCTION

Advanced multimedia and telecom applications exhibit very dynamic behavior. The number of tasks in the system can vary at run-time. Non-determinism is present due to e.g. user interaction. The application is expected to obey global timing constraints such as an expected frame rate in a real-time video or 3D rendering application. The computational requirements of such applications vary with orders of magnitudes between different image frames. Traditionally, a design that has to deliver a certain

E. Villar and J. Mermet (eds.), System Specification and Design Languages, 303–314.
© 2003 *Kluwer Academic Publishers.*

frame rate will be designed to guarantee the desired throughput even in the worst possible case. With the current dynamic applications this results in a design that is overdimensioned most of the time. Modern design methodologies (e.g. see [33]), however, can exploit the dynamic and non-deterministic behavior in such applications to trade-off the execution speed and power consumption on the platform. To enable such methodologies to fully exploit these properties of the applications, we present the MTG* model, a successor to both the Multi-Thread Graph (MTG) model [31] and the DF* model [6,26], and show how it highlights dynamism to allow more efficient analysis and exploration. Decisions about which parts of the specification will be implemented in hardware or software are not fixed in MTG*, since we do not want to unnecessarily constrain the optimization freedom in too early a stage of the design. We also introduce some extensions to MTG* to arrive at a Gray-Box model that allows to hide details of the specification that are unimportant for task level parallelism exploration.

The rest of the paper is organized as follows: in section 2 some related work will be discussed. Section 3 informally introduces the Gray-Box model using an example of a video application. Section 4 focuses on the constructs that are used to model the dynamic and non-deterministic behavior in the application. Section 5 touches some experiments we did starting from a gray-box model, and section 6 finally presents some conclusions.

2.   RELATED WORK

Many system specification and design languages exist. Many of these specialize in certain application domains. A good overview of models of computation underlying such languages can be found in [22].

For control-dominated designs, languages like StateCharts [24] and Esterel [3] can be used. These languages are typically not suitable for our target class of applications since the data-dominated signal processing kernels that are typically present in multimedia and telecom applications cannot be easily modeled.

Data flow and signal processing applications are usually modeled using some data flow paradigm: synchronous data flow [21], cyclo-static data flow [2], parameterizable data flow [1], dynamic data flow [10], Kahn process networks [17], YAPI [7]. Pure data flow systems are not suitable for our purposes, since they do not model non-deterministically occurring events in an elegant way, and usually do not allow dynamic task creation and deletion. YAPI [7] implements a 'select' operation which allows it to model non-deterministic events, but does not allow dynamic task creation. An advanced system like Ptolemy [20] uses heterogenous models of computation to allow the designer to choose the most suitable model for the problem at hand. It does allow dynamic task creation, but does not allow connecting a newly created task to a task that already exists in the system, because new ports cannot be created on top of an existing task.

For multimedia and telecom applications with both control-dominated and data-dominated application parts, models like SDL [13], SystemC [23], CoCentric System Studio [4], SpecC [9], SPI [15], FunState [30] or MTG [31] can be used. These approaches typically lack capabilities in one or more of the following domains: dynamic task creation and deletion (CoCentric Studio, MTG, SPI, FunState, SpecC, SystemC), non-deterministically occurring events (MTG) and modeling non-deterministic control flow (SpecC, SystemC). SDL [13] has evolved into a very complete modeling language, but for our purposes does not clearly enough separate dynamic and non-deterministic behavior from deterministic behavior. The dynamic and non-deterministic behavior can be buried in low levels of the model hierarchy, which complicates exploring large models. CoCentric Studio [4] does not support global shared variables. SPI [15] has no clear separation between control flow and data flow, because all control flow information is modeled using data carrying tokens. Extending SPI into FunState [30] remedies this situation, but still offers no support for dynamic task creation and deletion, or for dynamically managed large data types.

When looking at advanced multimedia and telecom applications with dynamic memory management and dynamic task creation, and non-deterministically occurring events, a language like e.g. C/C++ or Java [11,34] can be used. If on top of these requirements a good model for non-deterministic control flow is desired, and task level, loop level and instruction level parallelism must be made visible, the DF* model [6,26] can be used. Although both DF* and C/C++/Java can be used to describe advanced systems, they have the disadvantage of allowing points of non-determinism to become hidden arbitrarily deep inside the hierarchy. This makes handling of complex dynamic systems in terms of design exploration at the concurrent task level nearly impossible because of scalability issues. Moreover, these languages do not offer support for timing constraints and guarantees.

3.   INTRODUCTION TO MTG* AND GRAY BOX

3.1   A dynamic multimedia application

- Two incoming data streams are present. One is a digital data stream with up to 64 multiplexed video channels that are encoded in MPEG2. The other data stream is base band analog video input.
- There is one output screen onto which any number of incoming video channels can be displayed as selected by the user with a remote control.
- The user can indicate at any time that he/she is taking a break. The video recorder must then buffer the incoming video stream of which the sound channel is active, until the user resumes the viewing session. The other channels are not recorded, but displayed while the selected channel is buffered.

3.2   MTG* model

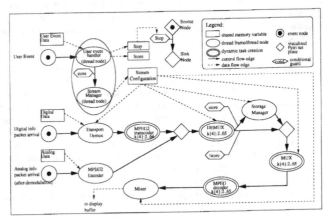

Figure 1. Task layer model of dynamic video recorder.

As described in the introduction one main idea behind the MTG* model is highlighting dynamic behavior in applications. To this end we model the application in two layers.

The first layer is the concurrent MTG* task[4] layer (see Figure 1), which contains thread frames bounded by dynamic constructs. Looking at the task layer model shows how the statically behaving parts of the application interact dynamically and shows the task level parallelism present in the system. MTG* thread frame boundaries show where a run-time scheduler may have to be activated. Reasons for activating the run-time scheduler include dynamically creating thread frames, or reconfiguring thread frames to accommodate the varying system load. As an example of thread frame creation, consider the MPEG2 Transcoder in Figure 1. The number of transcoder tasks that is created depends on the number of channels the user wants to watch at the same time. As an example of a thread frame reconfiguration, consider mapping the thread frame on a faster processor in the system. This results in a shorter execution time, but higher power consumption. In order to reduce run-time scheduling overhead, thread frame boundaries will be defined only at those places where it is really needed: thread frames that are dynamically created, event handling or non-determinism due to synchronization, semaphores or blocking reads. MTG* thread frames can be grouped hierarchically into MTG* tasks to improve the readability of large models. The concurrent task layer model behaves like a Petri net [28]. The Petri nets are extended because although they offer a natural way to model concurrency and non-determinism, they traditionally do not provide for easy modeling of data flow and dynamic task creation and deletion. Actual functionality is assigned to the Petri net transitions: an MTG* thread frame can be executed as soon as a token arrives on the input control port. The token is consumed when the execution finishes and a token is put on each outgoing control flow edge for which the associated guard condition is true. A Petri net place is implicitly present on each control flow edge. It is not visualized except when needed as explained in section 4.The tokens in the MTG* task layer model are valueless. The concurrent task layer thus clearly visualizes the dynamic behavior inside the application, as well as task level parallelism, important inter task data flow, inter task data dependencies and synchronization.

The second layer in the model is the intra task layer, which contains a hierarchical Control Data Flow Graph (CDFG) [25,32,18,16]. The MTG* intra task layer defines the functionality of the thread frames. Inside thread frames a hierarchical CDFG is present that visualizes intra-thread frame data flow, loop and instruction level parallelism, and intra-thread frame data dependencies. Figure 2 shows these two layers of abstraction schematically.

---

[4] Further on simply called task. Depending on the size and design of the application either a thread frame or a thread node will correspond to the intuitive notion of a task.

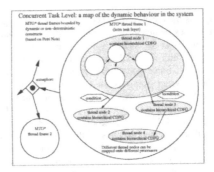

Figure 2. Levels of abstraction in MTG*/Gray-box

3.3   Gray-box model

Because the MTG* intra task layer contains fully detailed code, it can become quite large for large applications. We therefore introduced the notion of hidden and visible parts of the code. We do this by clustering parts of the CDFG in so-called thread nodes. A thread node is considered to be a black box. The code inside a thread node is said to be hidden. For large and complex systems, applying optimizing source code transformations on the full code of the application is not feasible because it typically contains too many details and the opportunities for improvement are not explicitly visible. By hiding well-chosen details, that are not relevant for the concurrent task-level analysis and exploration, in thread nodes, we obtain a Gray-Box model that remedies this situation. If needed during the design flow the thread nodes can be opened to access the underlying detailed code. Opening the thread nodes is called refinement[5].

These additional notions of hidden and visible code, and of hiding and refining thread nodes are what constitute the main differences between the MTG* model and the Gray-Box model: the MTG* model contains the fully detailed specification, whereas the Gray-Box model consists of an abstracted version of the specification. Using hiding and refinement as a way of managing complexity has the advantage that the global view of the system's dynamic behavior is not lost.

At this time, the boundaries of thread nodes are determined by heuristics based on the difference between minimum and maximum execution time

---

[5] The terminology refinement as used in this document has a different meaning than in other literature on formal system modelling.

(due to data dependencies), on the most important task-level data accesses (that should remain visible), and based on domain and data decomposition [8] of the thread frame. Guidelines have been defined to do this systematically but that is not the topic of this paper.

Different thread nodes inside one thread frame can be mapped to different processors on the platform to exploit the parallelism inside the thread frames. Thread nodes in the Gray-Box model roughly correspond to what most researchers would call a task, except that they are guaranteed to behave deterministically, and hence are amenable to design-time analysis.

A plethora of intra- and inter-task timing constraints and guarantees are added to the models [31]. These are not essentially different from the interval models in e.g. SPI [15], and so we will not discuss them further.

4.   DYNAMIC BEHAVIOR IN MTG*

In this section we focus on the constructs that MTG* offers to model dynamic and non-deterministic behavior. We will show which constructs are offered to model dynamic task creation, event with priority, semaphore, non-deterministic OR, non-deterministic choice and task-level synchronization.

The notations for these constructs are summarized in Figure 3. The ideas behind dynamic task creation in MTG* are partly inherited from the DF* model [6]. Like in the DF* model we use the idea of a configurator [26]. The configurator handles requests for dynamically instantiating, deleting or reconfiguring tasks, and makes sure that no inconsistencies occur. This configurator is part of the operating system running on the target platform. As in DF*, we use the geometrical parallelism notation [19] to concisely describe multiple tasks that execute identical functions in the system, and we use the same concept of multiports that allows specifying a varying number of data ports at run-time. As an example of this geometric parallelism notation, consider the MPEG2 transcoder thread frame in Figure 1. The inside notation $k\{4\}:2..64$ means that at this moment in time 4 tasks are active, and at any moment in time the number of tasks that are active in the system belongs to the interval [2,64].

An event is modeled by a construct that allows injecting a token from the environment into the model (the event node). In contrast to DF* no predefined event handling schemes are available. Instead, we have some primitive constructs that allow modeling all events we encountered in real-life applications so far.

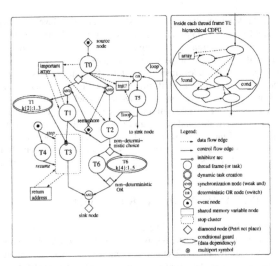

Figure 3. Overview of MTG* model elements

These constructs consist of: event node (receives token from outside world, e.g. a user pressing a button on a remote control), inhibitor arc (can be used to model priorities, see e.g. [5]) and a stop cluster. A stop cluster is a cluster of thread frames that are interrupted as soon as a token arrives on the stop cluster's control port. If desired, the address where the thread frame was interrupted can be exported and be used to have the thread frame resume. As is true for all model elements, a stop cluster has no predefined implementation, since this would introduce an implementation decision at too early a design stage. Note that events not necessarily occur non-deterministically, they could also occur periodically. This can be annotated on the model. In Figure 1 the arrival of information packets and the user interaction is modeled using event nodes.

Because the concurrent task layer model is an extension of Petri nets, we can model all non-deterministic control flow using a visualized Petri net place. This place is visualized using a diamond shaped symbol. Depending on the number of incoming/outgoing control flow edges connected to the diamond and the number of initial tokens, we can interpret it in a different way. Multiple incoming edges make for a non-deterministic OR behavior - non-deterministic because it is not known where a token will come from. Multiple outgoing edges connected to a diamond model a non-deterministic choice. Having both multiple incoming and outgoing edges with an initial token in the diamond allows to model semaphores. A diamond symbol with only incoming edges can be used as a sink node, and a diamond with only outgoing edges and an initial token can be used as a source node. Each MTG*/Gray-Box model has exactly one source and sink node. Figure 1 shows an example of a source and sink node and also of two non-deterministic OR nodes.

A situation in which a task can be started only when two other tasks have finished can be modeled using an AND node. The AND node can fire when each incoming control flow edge has an incoming token. As opposed to the non-deterministic OR node, also a deterministic OR node exists, which knows from which control flow edge to expect a token, depending on the value of an associated expression.

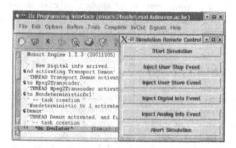

Figure 4. Screenshot of simulator running the video application.

5.   EXPERIMENTS AND RESULTS

We currently have developed a prototype Gray-Box simulator that performs simulation of Gray-Box models. Figure 4 shows a screenshot of the simulator running the example model is shown. At the moment we use an Xfig 3.2 [29] drawing of the MTG* task layer model as input, and generate executable Mozart/Oz3 code [27]. Oz3 is a multi-paradigm programming language designed for massive concurrency with built-in data flow synchronization mechanisms. It allows the programmer to combine techniques from data flow programming, higher order functional programming, logic programming, constraint satisfaction programming, object oriented programming and imperative programming at will. Because it can be extended with C or C++ code, it can be used as a very high-level glue language. The Oz3 code that is generated for a Gray-Box model carefully mimics the task layer model's semantics to simulate the concurrent task layer control and data flow including all dynamic and non-deterministic aspects. The intra-thread frame models can be simulated by linking in C/C++ code. We plan to extend this simulator to incorporate profiling and timing information for improved analysis purposes, and to make some provisions for automated (power consumption, time budget) trade-off curve extraction for concurrent task level exploration purposes.

The Gray-Box model described above is currently being used as input model for the Task Concurrency Management (TCM) methodology [33], which aims at exploiting the dynamic and non-deterministic behavior of multimedia and telecom applications to reduce the power consumption.

The first experiments using the TCM methodology starting from a Gray-Box model of a dynamic multimedia application have shown promising power consumption gains of about 50% as compared to a traditional real-time design approach.

6.   CONCLUSIONS

This paper shows how we approach modeling the dynamic aspects of advanced multimedia and telecom applications. The MTG* model we introduced can be extracted from a prototype implementation and serves as a starting point for systematic exploration of the design space. The model visualizes the task level parallelism and dynamic aspects of the system as clearly as possible and enables system design flows to take these into account.

The extensions of MTG* to a Gray-Box model allow us to selectively hide or show details of the specification that are needed during specific steps of a design flow. This enables the design flow to explore large and complex designs without loosing a global picture of the system behavior. To the best of our knowledge, the way we define the boundaries of MTG* thread frames to get the best possible view of dynamic behavior in an application is unique to MTG*. The concept of an MTG* thread frame bounded by dynamic constructs enables modern design methodologies to exploit the dynamic and non-deterministic behavior in the application to reduce e.g. power consumption.

ACKNOWLEDGEMENTS

This research is partially funded by a Philips grant. Geert Deconinck is a postdoctoral fellow of the Fund for Scientific Research (FWO). K.U.Leuven ESAT and IMEC are members of the DSP valley network.

7.   REFERENCES

[1]B. Battacharya and S. Battacharya. Parametrized dataflow modeling of DSP systems. In Proceedings of the International Conference on Acoustics, Speech, and Signal Processing, Istanbul, Turkey, June 2000.

[2]G. Bilsen, M. Engels, R. Lauwereins, and J. A. Peperstraete. Cyclo-static dataflow. IEEE Transactions on Signal Processing, 44(2):397--408, February 1996.

[3]F. Boussinot and R. de Simone. The ESTEREL language. In Proceedings of the IEEE, volume 79, pages 1293-1304, 1991.

[4]J. Buck and R. Vaidyanathan. Heterogeneous modeling and simulation of embedded systems in El Greco. In Proceedings of the 8th International Workshop on HW/SW Codesign (CODES), San Diego, California, May 2000.

[5]G. Chiola, S. Donatelli, and G. Franceschinis. Priorities, inhibitor arcs and concurrency in P/T nets. In Proceedings of the 12th International Conference on Application and Theory of Petri Nets, pages 182--205, Gjern, Denmark, June 1991.

[6]N. Cossement, R. Lauwereins, and F. Catthoor. DF-star: An extension of synchronous dataflow with data dependency and non-determinism. In Proceedings of the 3rd Forum on Design Languages FDL00/SSDL Workshop, Tübingen, Germany, 2000.

[7]E. de Kock, G. Essink, W.J.M. Smits, P. van der Wolf, J,-Y. Brunel, W. M. Kruijtzer, P. Lieverse, and K. A. Vissers. YAPI: Application modeling for signal processing systems. In Proceedings of the 37th Design Automation Conference (DAC00), pages 402--405, Los Angeles, CA, June 2000.

[8]I.Foster. Designing and building parallel programs: concepts and tools for parallel software engineering. Addison-Wesley, 1995.

[9]D. Gajski, J. Zhu, R. Domer, A. Gerstlauer, and S. Zhao. SpecC: Specification language and methodology. Kluwer Academic Publishers, 2000.

[10]S. Ha and E.A. Lee. Compile-time scheduling of dynamic constructs in dataflow program graphs. IEEE Transactions on Computers, 46, July 1997.

[11]R.Helaihel and K.Olukotun. Java as a specification language for hardware-software systems. In Proceedings of the International Conference on Computer-Aided Design, pages 690--697, November 1997.

[12]E. Hwang, F. Vahid, and Y. Hsu. Functional partitioning for reduced power. Technical Report CS-98-03, Department of Computer Science, University of California, Riverside, CA92521, May 1998.

[13]ITU-T Recommendation Z.100. Specification and description language (SDL), 1999.

[14]M. Jersak, D. Ziegenbein, and R. Ernst. A general approach to modeling system-level timing constraints. In Proceedings of the 4th Forum on Design Languages, Lyon, 2001.

[15]M. Jersak, D. Ziegenbein, F. Wolf, K. Richter, R. Ernst, F. Cieslok, J. Teich, K. Strehl, and L. Thiele. Embedded system design using the SPI workbench. In Proceedings of the 3rd Forum on Design Languages, September 2000.

[16] J.A.G. Jess. Tutorial G, high level modeling and formal verification, 1993. Electronic Design Automation Conference EDAC, Euro ASIC, Paris.

[17] G. Kahn. The semantics of a simple language for parallel programming. In Information Processing 74: Proceedings of IFIP Congress 74, pages 471--475, Stockholm, Sweden, August 1974. North-Holland.

[18] D. Lanneer. Design Models and Data-Path Mapping for Signal Processing Architectures. PhD thesis, Katholieke Universiteit Leuven, Leuven, Belgium, March 1993.

[19] R. Lauwereins, P. Wauters, M. Adé, and J. Peperstraete. Geometric parallelism and cyclo-static dataflow in grape-II. In Proceedings of the 5th International Workshop on Rapid Systems Prototyping, Grenoble, France, June 1994.

[20] E.A. Lee. Overview of the ptolemy project. Technical report, University of California, Berkeley, March 2001. Technical Memorandum UCB/ERL M01/11.

[21] E.A. Lee and D. G. Messerschmitt. Synchronous data flow. In Proceedings of the IEEE, volume~75, pages 1235--1245, September 1987.

[22] E.A. Lee and A. Sangionvanni-Vincentelli. A denotational framework for comparing models of computation. In Proceedings of the International Conference on Computer-Aided Design, 1996.

[23] S. Liao, S. Tjiang, and R. Gupta. An efficient implementation of reactivity for modeling hardware in the scenic design environment. In Proceedings of the 34th Design Automation Conference, Anaheim, California, June 1997.

[24] M. von der Beeck. A comparison of statecharts variants. In Proceedings Formal Techniques in Real Time and Fault Tolerant Systems, pages 128--148, 1994.

[25] M.C. McFarland, A.C. Parker, and R. Camposano. Tutorial on high-level synthesis. In Proceedings of the 25th ACM/IEEE Design Automation Conference, pages 330--336, San Francisco CA, June 1988.

[26] S. Mostert, N. Cossement, J. van Meerbergen, and R. Lauwereins. DF-star: Modeling dynamic process creation and events for interactive multimedia applications. In Proceedings of the 12th international workshop on Rapid System Prototyping, IEEE computer society, June 2001. Monterey, California, USA.

[27] Mozart Consortium. The mozart programming system, January 1999. Available at http://www.mozart-oz.org.

[28] J.L. Peterson. Petri Net Theory and the Modeling of Systems. Prentice-Hall, 1981.

[29] B.V. Smith. The xfig 3.2.3d user manual, may 2001. Available at http://www.xfig.org.

[30] L. Thiele, K. Strehl, D. Ziegenbein, R. Ernst, and J. Teich. Funstate - an internal design representation for codesign. In Proceedings of the ICCAD, IEEE/ACM International Conference on CAD, pages 558-565, San Jose, U.S.A., November 1999.

[31] F. Thoen and F. Catthoor. Modeling, Verification and Exploration of Task-Level Concurrency in Real-Time Embedded Systems. Kluwer Academic Publishers, Netherlands, first edition, 2000.

[32] T. Krol, J. van Meerbergen, C. Niessen, W. Smits, and J. Huisken. The sprite input language: an intermediate format for high-level synthesis. In Proceedings of the 3rd ACM/IEEE Europ. Design Automation Conference, pages 193-199, Brussels, Belgium, March 1992.

[33] P. Yang, C. Wong, P. Marchal, F. Catthoor, D. Desmet, D. Verkest, and R. Lauwereins. Energy-aware runtime scheduling for embedded multiprocessor SOCs. IEEE Design&Test of Computers, special ITC issue, 18(5):46--58, September-October 2001.

[34] J. Young, J. MacDonald, M. Shilman, A. Tabbara, P. Hilfinger, and R. Newton. Design and specification of embedded systems in JAVA using successive, formal refinement. In Proceedings of the Design Automation Conference, pages 70-75, June 1998.

Chapter 26

SPECIFICATION AND REFINEMENT OF HARDWARE COMPONENTS IN B

Stefan Hallerstede
KeesDA

Abstract: We use the B formalism to derive functionally correct synchronous circuits. To rep-resent the circuit we employ the hardware description language VHDL. This article outlines the development of a circuit design starting from an initial abstract functional specification of a system component. We discuss some topics involved in the translation to synthesisable VHDL and demonstrate the approach by way of an example.

1. INTRODUCTION

The traditional application domain of the B formalism [2] is software development. Recently, some work has been carried out to extend this domain into hardware development also [3, 14]. The B formalism is successfully used for formal specification and refinement of software systems. Refinement is a proof-based method to derive implementations from abstract initial specifications. Usually VHDL designs are verified by the creation of input stimuli, which are subsequently used in a simulation of the design. Using formal methods one achieves more dependable designs because their behaviour is verified by proof with respect to all permissible input stimuli. Model checking (e.g. [7]), theorem proving (e.g. [15, 17]) and combinations of both (e.g. [16]) have been applied successfully to solve such verification problems. State space explosion limits the application of model checking, whereas theorem proving produces huge proofs that are difficult to manage. In the combination of the techniques model checking is applied to sub-goals in a proof. This may yield significant productivity gains [16].

E. Villar and J. Mermet (eds.), System Specification and Design Languages, 315–325.

© 2003 Kluwer Academic Publishers.

However, this requires that those sub-goals are amenable to model checking, and it does not help to manage the complexity of proofs of complex systems. A strength of refinement is that it creates structured manageable proofs. Monolithic verification proofs are avoided [8]. From a broader perspective, the prospect of having a single modeling language for software and hardware at one's disposal is also very appealing. This could simplify the task of software/hardware co-design significantly.

In this paper we describe the principles of a translation of B machines into VHDL, which is one of the corner stones in our effort to apply B to hardware design. Descriptions of the refinement methodology of B are described in [3, 8] for example. We briefly describe the two approaches [3, 14] mentioned above. The approach taken by [14] is to use a particular B machine *PR* to express a register-transfer level design. The machine has one operation, which consists of a choice between multiple assignments depending on the state of the machine. A second machine *SM* is used to facilitate translation into synthesisable VHDL. It specifies the corresponding entity and architecture descriptions. Special variables *Clock*1 and *Clock*1 *EV ENT* are used to model clock synchronisation. The intention is to achieve a straightforward mapping to the corresponding VHDL description. The work presented in [3] is based on event-B [1]. The two major differences of [3] from [14] are that there is no explicit representation of a clock, and the final hardware description is not in VHDL. In [3] system specification is partitioned into two parts: an environment and a circuit. The two are executed alternately to model the interaction between the circuit and its environment. The circuit part of the system is refined towards a specific circuit design. Finally it is translated into a block diagram, which represents a synchronous design. In both approaches [3] and [14] only data-types close to hardware are being used on the implementation level. In B Modelling real-time properties is not supported. However, it is well suited for modeling synchronous systems where we do not need to express real-time properties explicitly. This paper presents an approach to the derivation of a circuit from a B machine. As done in [14] we use synthesisable VHDL to represent the synchronous hardware component that corresponds to a B machine. Similar to [3] we do not have an explicit clock in B. It is introduced during the translation into VHDL together with necessary timing constraints. The translation suggested in this article does assume that a B machine to be translated has only one operation. We have extended the translation to deal with multiple operations. But this is not presented here. In fact, the translation [13] itself does not target VHDL directly but a hardware semantics described in predicate logic. This translation can subsequently be used with any hardware description language that has a suitable semantics.

One aim of our work is to define a subset BHDL of statements and data-types of B for hardware synthesis. This has been successfully done for software resulting in the definition of B0. Another important aim is to describe the formal semantics of a class of synthesisable VHDL designs using B machines. The reasoning about correctness of complex designs could then be carried out entirely in B. Doing this in VHDL itself tends to be rather difficult because of the complex semantics of the language [6].

1.1　Description of a Storage Element in B

In this section we present the specification of a storage element in B. In the following we refer to the storage element as "memory". Machine *MEMORY* is well suited to discuss many fundamental issues of the translation into VHDL. However, some complications that arise in the translation do not show with this simple machine. They will be dealt with in the context of machine *BUFFER* later. The initial specification is the B machine shown in Figure 1. The corresponding VHDL design is presented in section 3. The memory stores integer numbers ranging from 0 to *width*, and has a maximal storage capacity of *cap*. The set *ADDR* specifies the range of valid addresses of the memory. Constants *cap* and *width* serve as machine parameters which are to be instantiated in an implementation of *MEMORY*.

The initialisation of machine *MEMORY* (Figure 1) sets all values to zero. The operation *mem_op* of machine *MEMORY* has three functionalities: the memory can be reset, a value can be written into the memory or read from the memory. The functionalities to reset or write must be activated explicitly by setting the corresponding parameter of *mem_op* to TRUE.

```
MACHINE MEMORY
CONCRETE CONSTANTS
cap, width, WORD, ADDR
PROPERTIES
cap ∈ N1 ∧ width ∈ N1 ∧ WORD = 0 .. width ∧ ADDR = 0 .. cap-1 ;
VARIABLES mem
INVARIANT mem ∈ ADDR → WORD
INITIALISATION mem := ADDR × {0}
OPERATIONS
mout ← mem_op(mad, minp, mreset, mwrite) =
pre mad ∈ ADDR ∧ minp ∈ WORD ∧ mreset ∈ BOOL ∧ mwrite ∈ BOOL then
if mreset = TRUE then mem := ADDR × {0}
elsif mwrite = TRUE then mem(mad) := minp
end || mout := mem(mad)
end
END
```

Figure 1: Abstract Memory Machine

They cannot be activated at the same time. The value *mem*(*mad*) read at address *mad* is always returned in the output parameter *mout*.

Machine *MEMORY* is refined into implementation *STORAGE* which is shown in Figure 2. The constants and variables have been tagged "concrete" to indicate that they are implementable. This is supported by the B formalism. When concrete variables and constants are introduced in intermediate refinement steps, it is enforced that they are not refined further. They are considered to be part of the implementation. The values of the constants are specified in the VALUES clause. Note that the type of variable *mem* is given implicitly by the variable of the same name in machine *MEMORY*

2.   VHDL-IMPLEMENTATION OF THE STORAGE ELEMENT

Implementation *STORAGE* (Figure 2 below) of machine *MEMORY* can be translated into VHDL as outlined in this section. The translation follows the structure of implementation *STORAGE*. We only present the principles of the translation and not a formal proof. The proof is based on a semantics modelling concurrency in VHDL as described in [12].

```
IMPLEMENTATION STORAGE
REFINES MEMORY
VALUES cap = 4; width = 8; WORD = 0 .. width; ADDR = 0 .. cap-1 ;
CONCRETE VARIABLES mem
INITIALISATION mem := ADDR × {0}
OPERATIONS
out ← mem_op(mad, minp, mreset, mwrite) =
begin
if mreset = TRUE then mem := ADDR × {0}
   elsif mwrite = TRUE then mem(mad) := minp
   end ‖ mout := mem(mad)
end
END
```

Figure 2: Implementation of the Memory Machine

The items declared in the constants section of machine *STORAGE* are translated into a package which is used by all implementations importing machine *STORAGE* and, of course, machine *STORAGE* itself. The package is presented in Figure 3. The values of the constants cap and width are taken from the values clause of *STORAGE*. Figure 4 shows the interface ofthe memory design, entity storage. The input and output ports correspond to the respective input and output parameters of operation *mem_op*. The clock input port has been introduced to implement a register that stores the state of the design. In synthesisable VHDL a circuit is subdivided into two disjoint parts: a synchronised circuit and a combinatorial circuit. The synchronised circuit is driven by the clock and models a register which implements the variables of an implementation. The combinatorial circuit is independent of the clock. It models a state transition corresponding to the execution of the operation of an implementation. Figure 5 shows the architecture mixed specifying the behaviour of entity storage. The synchronised part of the design is represented by process sync, and the combinatorial part by process comb. The before and after values of variable *mem* have been named into mem_pre and mem_post, respectively. This is based on the representation of operation *mem_op* as a before-after predicate as defined in [2, 11]. The signals mreset and mwrite are used to activate particular parts of the circuit. Their presence in B, thus, makes the control of the circuit corresponding to machine *MEMORY* explicit.

```
package storage param is
constant cap : NATURAL := 4;  width : NATURAL := 8;
subtype WORD is NATURAL range 0 to width; ADDR is NATURAL range 0 to
cap-1;
end package storage param;
```
Figure 3: Package storage param

```
entity storage is port( clock : in BIT;
mout : out WORD; minp : in WORD; mad : in ADDR;
mreset : in BOOLEAN; mwrite : in BOOLEAN);
end entity storage;
```
Figure 4: Entity storage

On a rising edge of the clock, i.e. clock'EVENT and clock = '1', process sync stores the after value mem_post of *mem* in the before value mem_pre. Using mem_pre as the current state, the successor state is computed by process comb during the remainder of the clock cycle, i.e. till the next rising edge of the clock. The behaviour of the circuit would be ill defined if signals clock and mreset, say, were to change at the same time. The order in which processes sync and comb were to be executed would not be deterministic. To solve possible timing problems the following constraints are required to be satisfied by any environment of a synthesisable VHDL design [18]:

1. it must be ensured that all signals are stable on the rising clock edge; and

2. input signals are only changed some time after the rising clock edge.

Constraint 1 ensures that the state being stored by a register is really the after state. Constraint 2 solves the problem of the activation order of the two processes above. Note that the first constraint emerges when the instantaneous evaluation of operation *mem_op* is replaced by the execution of a circuit consisting of gates where physical signals are propagated. The physical signals on the gate level should not be confused with the (abstract) signals modelled on the VHDL level as represented in the predicate.

```
architecture mixed of storage is
```

```
signal mem_pre;mem_post : array ADDR of WORD;
begin
    sync : process (clock) is begin
    if clock'EVENT and clock = '1' then mem pre<=mem_post; end if;
    end process;
    comb : process (mem pre, minp, mad, mwrite, mreset) is begin
    mem post<=mem_pre;
    if mreset = TRUE then mem_post<=(others => 0);
    elsif mwrite = TRUE then mem_post(mad) <= minp; end if;
    mout <= mem_pre(mad);
    end process;
end architecture mixed;
```

Figure 5: Architecture of Entity storage

3.   DESCRIPTION OF A FIRST-IN-FIRST-OUT BUFFER IN B

To demonstrate our approach on a more complex example we use a FIFO buffer. Especially sequential composition and operation call are treated which do not occur in implementation *STORAGE* presented in section 2. Figure 6 shows the initial specification of the buffer. Its storage capacity is given by *cap*, the word size by *width* and its state described by the four variables *queue; sz; fst; lst*. Values inserted into the buffer are stored in variable *queue*. Variables *fst* and *lst* are the indices of the first and the last element in the buffer. Finally variable *sz* contains the number of elements in the buffer.

The machine has one operation *buf_op* that makes functionality available to reset the buffer, to put a value *winp* into the buffer, to get a value *wout* that is stored in the buffer, and to inquire whether the buffer is full or empty. Most of the used mathematical notation should be clear. In operation *put* we use the relational overwrite operator: *queue* is assigned value *ww* at position *xx* and left unchanged everywhere else. The three parameters *reset*, *put*, and *get* give access to the corresponding functionality of operation *buf_op*. It is also required that the buffer must not be full when a value is put into the buffer, i.e. *put* = TRUE $\Rightarrow$ *sz* < *cap*. The two inquiries whether the buffer is full or empty are performed unconditionally.

Machine *BUFFER* of Figure 6 is not suitable for translation into VHDL. The variables *queue*, *fst*, and *lst* have infinite types. Using data-refinement we change the state representation so that it contains only variables of finite type. Figure 7 shows the final implementation of machine *BUFFER*, called *FIFO*. It imports machine *MEMORY* (see Figure 1) for storage and retrieval of values. We use machine *MEMORY* to demonstrate how operation calls in B can be translated into VHDL. The machine parameters in the constants The implementation of machine *BUFFER* declares three own variables *null, prem, dern*. Variable *null* signals that the buffer is empty, or not. Variables *prem* and *dern* contain the indices of the first and last element in the buffer respectively. There is a single call of operation *mem_op* of machine *MEMORY* in the body of *buf_op*. This is necessary because, when translated into VHDL, *mem_op* models a physical entity with the corresponding functionality. Multiple calls to *mem_op* in B would correspond to multiple copies of this physical entity in VHDL.

MACHINE *BUFFER*
ABSTRACT CONSTANTS *cap, width*
PROPERTIES
$cap \in N1 \wedge width \in N1$
VARIABLES *queue; sz; fst; lst*
INVARIANT
$fst \in N1 \wedge lst \in N \wedge fst\text{-}1 \le lst \wedge queue \in 1 .. lst \rightarrow 0 .. width \wedge sz \in N \wedge sz = card(fst .. lst)$
INITIALISATION
$fst, lst, sz, queue := 1, 0, 0, \varnothing$;
OPERATIONS
wout, full, empty $\leftarrow$ *buf_op(winp, reset, put, get)* =
pre $winp \in 0 .. width \wedge reset \in BOOL \wedge put \in BOOL \wedge get \in BOOL$
then
    if *reset* = TRUE then *fst, lst, sz, queue* := $1, 0, 0, \varnothing$
    elsif *put* = TRUE then *lst, queue, sz* := $lst + 1$, $queue <+\{ lst + 1 \mapsto winp \}$, $sz + 1$
    elsif *get* = TRUE then *wout, fst, sz* := $queue(fst), fst + 1, sz\text{-}1$
    end $\|$ *full* := $bool(sz = cap)$; *empty* := $bool(sz = 0)$
end;
END

Figure 6: Abstract FIFO Buffer Machine

IMPLEMENTATION *FIFO*
REFINES *BUFFER*
IMPORTS *MEMORY*
CONCRETE VARIABLES *null; prem; dern*
INVARIANT $null \in BOOL \wedge prem \in ADDR \wedge dern \in ADDR$
INITIALISATION *prem, dern, null* := 0, 0, TRUE
OPERATIONS
wout, full, empty $\leftarrow$ *buf_op(winp, reset, put, get)* =
var *ad* in
if *put* = TRUE then *ad* := *dern* else *ad* := *prem* end;
wout $\leftarrow$ *mem_op(ad, winp, reset, put)*;
if *reset* = TRUE then *prem, dern, null* := 0, 0, TRUE
elsif *put* = TRUE then
$null$:= FALSE $\|$ if *dern* = *cap* - 1 then *dern* := 0 elsc *dern* := *dern* + 1 end

elsif *get* = TRUE then
if *prem* = *cap*-1 then *prem*:=0 else *prem* := *prem*+1 end; *null*:=bool(*pre* =*dern*)
end || *full* := bool(*prem* = *dern* ∧ *null* = FALSE) || *empty* := bool(*null* = TRUE)
end;
END

Figure 7: Concrete FIFO Buffer Machine

4.   VHDL-IMPLEMENTATION OF THE FIFO BUFFER

The principle of the translation of implementation *FIFO* is similar to that of implementation *STORAGE*. However, to deal with the operation call and with the sequential composition in operation *buf_op* of *FIFO* we split the combinatorial part of the design into several processes (and blocks). Because *FIFO* imports machine *MEMORY* , constants declared in *MEMORY* are visible in *FIFO*. In the resulting VHDL the design fifo corresponding to *FIFO* uses package storage param. Note that *FIFO* imports machine *MEMORY* but in the translation *MEMORY* is replaced by its implementation *STORAGE*, of course. Note also, that to simplify the presentation we have chosen all names in the B specifications so that renaming is avoided.

Figure 8 shows the entity description resulting from the translation of implementation *FIFO*. Signal clock is used for synchronisation. The remaining signals correspond to the parameters of operation *buf_op*. The architecture of entity fifo is shown in Figure 9. As in the case of the storage design we introduce two signals v_pre and v_post for each variable *v* of machine *FIFO*. A process sync is defined in the architecture of fifo to implement the state transition. The architecture also declares local signals for the communication with design storage, i.e. signals for the parameters of operation *mem_op* of machine *STORAGE*. Design storage is imported into the architecture as a component comp. The local signals and the clock signal are mapped to the signals with the same name in the "port map". This makes design storage available in the architecture of fifo as if it was a block declared locally under the name comp.

entity fifo is port(clock : in BIT;
wout : out WORD; winp : in WORD;
full : out BOOLEAN; empty : out BOOLEAN;
reset : in BOOLEAN; put : in BOOLEAN; get : in BOOLEAN);
end entity fifo;

Figure 8: Entity fifo

architecture mixed of fifo is
signal prem_pre; prem post,dern_pre, dern_post, null_pre, null_post : ADDR;

```
signal mout, minp : WORD; signal mad : ADDR; signal mwrite, mreset :
BOOLEAN;
begin
comp : entity storage(mixed) port map (clock, mout, minp, mad, mreset, mwrite);
sync : process (clock) is begin if clock'EVENT and clock = '1' then ... end if; end
process;
comb : block ... -- see figure 10
end architecture mixed;
```

Figure 9: Architecture of entity fifo

The combinatorial part of design fifo is modelled as a block comb and
not as a process like that of design storage. It is shown in Figure 10. We
discuss the translation of operation *buf_op* into block comb in the following
paragraphs. The presentation follows the structure of the operation.

Body of the Operation. The local variable *ad* of the var statement in the
body of operation *buf_op* is translated into a local signal *ad_loc* of block
comb. A local variable in a B implementation does not have a before-value,
i.e. *ad_loc* corresponds to an after-value. We have chosen a different
naming scheme for local variables to distinguish more easily them from
machine variables.

Body of the var Statement. The body of the var statement consists of a
sequence of three statements if ... end ; call of *mem_op* ; if ... end executed
in parallel with the assignments the output parameters *full* and *empty*. The
three statements change different sets of variables. They can be executed
concurrently in VHDL. References to a variables in an earlier statement in
the sequence correspond to signal propagations.

First if Statement. The first if statement is translated into a
corresponding process if1 which assigns a value to signal ad_loc.

Operation Call. The call of operation *mem_op* is translated into a block
call containing a set of concurrent signal assignments. Note that these signal
assignments are concurrent, as opposed to the sequential signal assignments
in process bodies. Concurrent signal assignments may only assign distinct
signals. The assignments connect the input and output ports of design
storage to signals of design fifo as specified by the call.

Second if Statement. The second if statement is translated into a block
if2 which is decomposed into further blocks and processes. The variables
changed by three branches of the if statement are not disjoint. This is dealt
with in the following way: Each branch br1, br2, br3 changes is own local
copy of the after-value of the variable. The branches are executed
concurrently and the correct after-value is chosen by a driver process for
each variable changed in one of the branches.

```
comb : block signal ad_loc : ADDR; begin
if1 : process (put, dern_pre, prem_pre) is begin
if put = TRUE then ad loc<= dern_pre else ad loc <= prem_pre end if;
end process;
```

```
call : block begin
wout<=mout; minp<=winp; mad<=ad_loc; mreset<=reset ; mwrite<=put;
end block;
if2 : block signal
br1_prem_post,br3_prem_post,br1_null_post,br3_null_post:ADDR;
signal … -- further signal declarations
begin
br1 : block begin -- branch "reset = TRUE"
br1_prem_post<=0; br1_dern_post<=0; br1_null_post<=FALSE;
end block;
br2 : block … -- branch "put = TRUE"
br3 : block … -- branch "get = TRUE"
prem_driver :  process … -- choose correct post-values
dern_driver : process … null_driver : process …
end block;
full<=(prem_pre = dern_pre and null = FALSE) ; empty<=(null = TRUE);
end block;
```

Figure 10: Combinatorial part of design fifo

5.   CONCLUSION

We have demonstrated a method to translate B machines into VHDL designs by means of examples. The approach poses fewer structural constraints on the B machines compared to the approach of [14]. Our present suggestion for the subset BHDL of B for hardware modelling consists of multiple assignments, sequential and parallel composition, conditional statements, for loops (not presented in this article), and types that correspond readily to types of synthesisable VHDL. One of our next steps is to extend the translation so that it can deal with a larger class of BHDL machines. In particular, explicit intermediate states in the VHDL design are required to deal with multiple calls to operations of a machine. The number of these states should be kept small to avoid wait cycles where possible.

Using B machines as an abstract representation of VHDL designs, reasoning about their properties can be carried out in B. This applies especially to the refinement of designs. It is less complex to do this in B than doing it in VHDL directly, because one does not have to deal with the intricacies of VHDL semantics. Additionally, the use of B for hardware design opens up the possibility of having a single formalism to model both hardware and software components. This could make it possible to use B as a basis for software/hardware co-design.

ACKNOWLEDGMENT

The work presented was done at KeesDA with the support of Jean Mermet. Adam Morawiec and Michael Butler helped to improve the quality of this article. Discussions at ClearSy, a member of the PUSSEE project within which this work is conducted, have helped to clarify the relationship between B and VHDL.

6. REFERENCES

[1] J.-R. Abrial and L. Mussat. Introducing Dynamic Constraints in B. In D. Bert, editor, *B'98: Recent Advances in the Development and Use of the B-Method*, volume 1393 of *LNCS*, pages 83–128, 1998.

[2] Jean-Raymond Abrial. *The B-Book – Assigning programs to meanings*. Cambridge University Press, 1996.

[3] Jean-Raymond Abrial. Event Driven Electronic Circuit Construction. Unpublished, August 2001.

[4] Peter J. Ashenden. *The Designer's Guide to VHDL*. Morgan Kaufmann, 1996.

[5] B-Core (UK), Harwell, United Kingdom. *B-Toolkit*. Software.

[6] P. T. Breuer, C. D. Kloos, A. M. Lopez, and N. M. Madrid. A Refinement Calculus for the Synthesis of Verified Hardware Descriptions in VHDL. *ACM TOPLAS*, 19(4),1997.

[7] Jerry R. Burch, Edmund M. Clarke, David E. Long, Kenneth L. McMillan, and David L. Dill. Symbolic Model Checking for Sequential Circuit Verification. *IEEE Transactions on Computer-Aided Design of Integrated Circuits and Systems*, pages 401–424, 1994.

[8] Dominique Cansell, Ganesh Gopalakrishnan, Mike Jones, Dominique Méry, and Airy Weinzoepflen. Incremental Proof of the Producer/Consumer Property for the PCI Protocol. In D. Bert, J. P. Bowen, M. C. Henson, and K. Robinson, editors, *ZB 2002: Formal Specification and Development in Z and B*, volume 2272 of *LNCS*, pages 22 – 41, 2002.

[9] ClearSy – Systems Engineering, Aix-en-Provence, France. *Atelier B*. Software.

[10] ClearSy – Systems Engineering, Aix-en-Provence, France. *Manuel de Référence du Langage B*, 2002.

[11] Steve Dunne. A Theory of Generalised Substitutions. In D. Bert, J. P. Bowen, M. C. Henson, and K. Robinson, editors, *ZB 2002: Formal Specification and Development in Z and B*, volume 2272 of *LNCS*, pages 270–290, 2002.

[12] Max Fuchs and Michael Mendler. A Functional Semantics for Delta-Delay VHDL Based on FOCUS. In C. D. Kloos and P. T. Breuer, editors, *Formal Semantics for VHDL*, pages 9–42. Kluwer Academic Publishers, 1995.

[13] Stefan Hallerstede. B for Hardware Specification. to be published.

[14] Wilson Ifill, Ib Sorensen, and Steve Schneider. The Use of B to Specify, Design and Verify Hardware. Unpublished.

[15] Laurence Pierre. Describing and Verifying Synchronous Circuits with the Boyer-Moore Theorem Prover. In P. Camurati and H. Eveking, editors, *Correct Hardware Design and Verification Methods*, volume 987, 1995.

[16] K. Schneider and T. Kropf. Verifying hardware correctness by combining theorem proving and model checking. In J. Alves-Foss, editor, *Higher Order Logic Theorem Proving and Its Applications: Short Presentations*, pages 89–104, 1995.

[17] John P. Van Tassel. An Operational Semantics for a Subset of VHDL. In C. D. Kloos and P. T. Breuer, editors, *Formal Semantics for VHDL*, pages 9–42. Kluwer Academic Publishers, 1995.

[18] VSI Working Group. IEEE P1076.6/D2.01 – Draft Standard For VHDL Register Transfer Level Synthesis. Unapproved Draft, IEEE, 2001.

Chapter 27

ENSURING SEMANTIC INTEGRITY IN KNOWLEDGE BASES FOR EMBEDDED SYSTEMS

Dieter Monjau, Mathias Sporer
Chemnitz University of Technology, Department of Computer Science, D-09107 Chemnitz, Germany, Strasse der Nationen 62

{dmo,masp}@informatik.tu-chemnitz.de

Abstract: The design process of embedded systems is supported by a large number of tools. The more domains are included in the embedded system (optic, mechanic, electronic, and so on) the more different tools must exchange their information. In order to reuse components the knowledge engineer must be able to construct knowledge bases suitable for special requirements. From the data processing point of view, the way of specification down to implementation is tantamount to the increasing amount of new information which must be consistent with existing information. In this paper we propose a new approach regarding the construction and maintenance of databases based on an extensible algebra. In this way, the customer himself can build up a database which is adapted optimally to a specific domain. The focus lies on the preservation of the semantic integrity which is secured independently of the data model and accessing tools.

Key words: embedded systems, knowledge base, semantic integrity

1. INTRODUCTION

Today's design of embedded systems is characterized by an increasing number of components such as processing units (microprocessors, ASIC,

E. Villar and J. Mermet (eds.), System Specification and Design Languages, 327–340.
© 2003 *Kluwer Academic Publishers.*

FPGA) and communication units (busses, shared memories), but also by mechanical and optical devices, and various kinds of software. However, these new capabilities cannot be used completely because appropriate design tools are missing. In the past, a lot of good methodologies where proposed for special domains or for a special level of abstraction in the design process [5][8]. The difficulty consists in completely supporting the transformation of draft knowledge from a specification given in informal manner up to code generation and/or layout [10]. In our opinion the designer himself must be able to configure an environment suitable for the special task ahead. We present a common methodology for the design process of embedded systems based on a flexible and extensible meta model. The meta model not only covers the structure of all pieces of information which must be managed in the design process, it also provides mechanisms to simulate the system in a very early stage of the draft. Commercially available tools can be integrated into our approach. In this case, the focus lies on the preservation of the semantic integrity of all data (also regarding reused components). The new capabilities of active database systems are in use to achieve this goal.

2.   DESIGN PROCESS OF EMBEDDED SYSTEMS

The common design process of embedded systems can be implemented by a model shown in Figure 1 (similar to the waterfall model). In the first step the designer collects all functional and non-functional requirements of the system to be sketched. Functional requirements determine all the services which the customer using embedded systems expects. Non-functional requirements like response time, power consumption, reliability, overall costs etc. are properties of the system. These must be fulfilled according to the customer's order.

The system designer builds a specification model from these requirements. He can use the UML modeling technology to define *use cases* for the system. The specification model must then be mapped onto a function model consisting of implementation-independent descriptions of the behavior of subfunctions. An important demand of semantic integrity consists in the fact that the entirety of all subfunctions shows the same behavior as the model of specification. The next design step determines the architecture needed to implement these functions. In the architecture model processing functions carried out on the same processor are grouped forming so called architecture blocks. The last phase of the design process provides the implementation of the architecture blocks. This can be done by behavior descriptions written in a hardware description language like VHDL or - regarding software - in imperative programming languages.

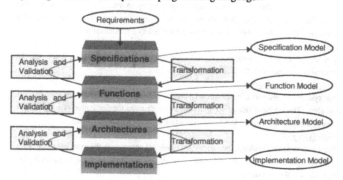

Figure 1: Synthesis steps in the common draft methodology

3.   A KNOWLEDGE BASE

In each phase of the design process information must be interchanged between different tools, i.e. the same information must be available to different kinds of representation to allow foreign tools to process it. For this reason, we use a DBMS for the administration of design knowledge. The knowledge base consists of four logical layers in accordance with the stages of the design process depicted in Figure 1. In each layer hierarchically ordered graphs are used to represent the design knowledge. The set of nodes of the graphs is formed by so called knowledge units. The set of edges is formed by relations between them. A knowledge unit is a class definition similar to OOPL extended by the following features [1][4]:

a set of attributes

For each attribute the knowledge engineer has to provide a unique name, a data type, an interval of validity (optional), and some methods for calculating the attribute's value.

a set of port definitions

Ports describe the points of communication between knowledge units inside the knowledge base as well as between the components in the real system. Ports are defined by their name, their data type, and their direction of transmission.

a set of subclasses

In order to express the aggregation/decomposition relation, each class can have several subclasses. The subclass is connected to the superclass by an inner layer relation of type *has parts*. The edges of these relations are named. A subclass can occur many times per superclass. The cardinality of the relation is determined by a special variable.

a task-graph

If a special execution sequence of the subfunctions is required, a task-graph can be established by describing the probability of transition.

a set of structure equations

Our system defines the communication between classes in an abstract way by structure equations. The left hand side of such an equation determines the destination of a signal. The term on the right hand side refers to a class that supplies the data or constant.

a set of views

When knowledge units should be mapped on a physical component, the exact behavior of the target component must be known. Therefore, the source texts of these components are maintained in the knowledge base.

a set of relations to the architectures

Since often a wide range of architectures can be built for the same system, the knowledge engineer determines by inter-layer relations of type *architector* which classes can correspond with special parts of the given architecture.

a set of relations to the implementations

Also for mapping an architecture of subfunctions onto the implementation layer, more than one component is available.

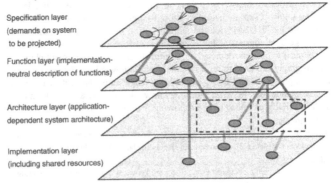

Specification layer
(demands on system
to be projected)

Function layer (implementation-
neutral description of functions)

Architecture layer (application-
dependent system architecture)

Implementation layer
(including shared resources)

Figure 2: Logical layers in the knowledge base

4.   SEMANTIC INTEGRITY

At the first configuration level of our system, the constituents of a knowledge unit are formulated by the descriptive language CLINT++ (Class and Instance Notation) [11]. To illustrate the usage of selected constituents mentioned above, we will give an example [6][7]. The cooperation of functions might be modeled as follows:

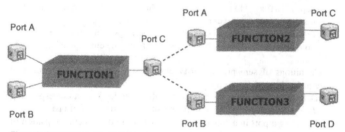

Figure 3: Cooperation of functions

The algorithm of function 1 decides whether function 2 or function 3 should be supplied with data produced by function 1. The probability of the transition from function 1 to function 2 is 90 per cent. Ports A and B of function 1 receive their data from the system's environment. Port C of function 2 and Port D of function 3 send their data to the superordinate system. CLINT++ describes the structure of this system and its internal communication (dashed lines in Figure 3) in textual manner as follows:

```
function class TOP is            function class FUNCTION1 is
parts (-- owns 3 subfunctions    port (
   step1: 1 FUNCTION1;              Port_A: in  natural;
   step2: 1 FUNCTION2;              Port_B: in  natural;
   step3: 1 FUNCTION3               Port_C: out integer);
   );                            end FUNCTION1;
tasks (-- edges of the task-graph function class FUNCTION2 is
   (#source, step1, 1),          port (
   (step1, step2, 0.9),            Port_A: in  natural;
   (step1, step3, 0.1),            Port_C: out integer);
   (step2, #sink, 1),            end FUNCTION2;
   (step3, #sink, 1));           function class FUNCTION3 is
 structure (-- communication     port (
   step2.Port_A := step1.Port_C;   Port_B: in  integer;
   step3.Port_B := step1.Port_C);  Port_D: out natural);
end TOP;                         end FUNCTION3;
```

Conditions of semantic integrity must be fulfilled when the generated system should work correctly:

- Deleting or updating the names of subclasses leads to a corresponding *delete-* or *update*-operation in the set of parts and the set of tasks.
- Updating a port in a subclass causes changes in the structure equation of the superclass if one of the equation's terms refers to this port.
- Deleting a port in a subclass causes the deletion of all structure equations containing a reference to this port.
- The attempt to change a port's property (e.g. update of transmission direction) must be suppressed by the design system if the signal assignment becomes impossible.
- A warning must be generated if not all values of the sending port can be processed by the receiving port (This can happen in the first structure equation when function 1 sends a negative value which cannot be understood by function 2 since port A is defined as natural.).

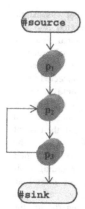

Figure 4: Deadlock checking

Another integrity problem, which is not solvable by means of conventional DBMS, is the recognition of deadlocks during the process of sketching the system.

Already in the design phase, connections between functions which can lead to deadlocks at runtime should be recognized by the knowledge based system.

In the task-graph shown in Figure 4 such a deadlock occurs. The activity condition of function p_2 requires a token provided by p_1 <u>and</u> a token provided by p_3. The function p_3 can start when p_2 has finished – the system is running in deadlock state.

The task-graph can be described as stated below (CLINT language):

These few examples already show that many different relationships can exist between the knowledge units. Some of the dependencies can be added to the database by the knowledge engineer, others are valid only in a special period of the development [3].

Commercially available DBMSs provide mechanisms to ensure the logical integrity of the stored data. Logical integrity is usually defined as a subset of semantic integrity. The intensity of checking the logic integrity depends on the data model: in RDBMS, keys and reference integrity between tables are checked by the system; in OOBDMS, the object structure captures some of the semantic conditions.

Specific integrity conditions are still left behind which the customer must provide compliance for. In the past they were expressed by triggers in the front end of a database application. This approach fails if a higher number of applications is allowed to perform database-write-operations. Figure 5 shows a typical scenario for using a knowledge base in the design process.

```
function class P is
  comment("possible deadlocks";
  "in iterative problems");
  parts (
    step1: 1 P1;
    step2: 1 P2;
    step3: 1 P3);
  tasks (
    (#source, step1, 1),
    (step1,   step2, 1),
    (step2,   step3, 1),
    (step3,   step2, 0.8),
    (step3,   #sink, 0.2));
  structure (
    step2.B = step1.A;
    step2.C = step3.F;
    step3.E = step2.D);
  activity (
    step2: B > 0 && C > 0);
end P;
```

Each condition of semantic integrity not supported by any mechanisms of logical integrity must be covered by tools such as shown in Figure 5. Consequently, all tools which are connected to the knowledge base would have to consider a change of the integrity rules at the same time. Such a demand is unrealistic in the case of commercial tools since the customer cannot intervene in their source code. Even regarding tools the customer developed himself, an automated checking of their correctness with regard to the integrity-rules is difficult because different languages and programming paradigms are used in the state of evolution concerned [2].

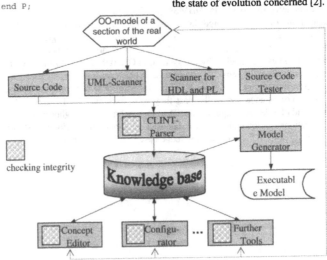

Figure 5: Temporary aspects of semantic integrity

To overcome these problems, all semantic rules must be shifted into the responsibility of the DBMS.

5.   ACTIVE DATABASES

Conventional DBMSs react at orders which an inquiry language sets up. In contrast, active databases can detect events caused by accesses to stored objects. Additionally, events from the operating system or even from foreign systems can be recognized. This technology is orthogonal to the data model, i.e. active mechanisms can be included in relational as well as in object-oriented systems. The comparison between currently known strategies for the activation of integrity-checking is shown in Figure 6.

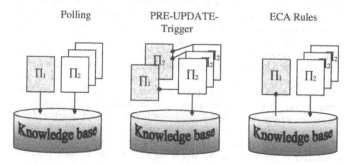

Figure 6: Activation strategies for integrity-checking

Polling
If processes $\Pi_2 \dots \Pi_n$ can submit write-operations, then a process Π_1 must be activated cyclically in order to check the semantic integrity of the database. Since no information about the last modifying operation is available when Π_1 becomes active, the whole database must be checked for inconsistencies.
Trigger
The disadvantages of the polling strategy are eliminated if a trigger is used to test the effects of the transaction before every write-operation. The trigger can terminate the calling transaction without "commit" before the database runs into an inconsistent state. The same functionality must be implemented in all applications.
ECA Rules

ECA rules (Event, Condition, Action) are an essential component of active databases. The insertion of a data record into a table or the deletion of an object are examples of events which the database registers. After the event has occurred, a condition is evaluated. The result of this evaluation decides whether the specified action will be carried out. The advantage of this procedure lies in its independence of the calling transaction, i.e. applications added to the design system later have no influence on the formulation of the integrity rules.

6.   PROCESSING OF ECA-RULES

DBMSs available today still do not support comprehensive processing of ECA-rules. By simulating the events using special triggers, we proved this comprehensive processing of ECA-rules to be successful regarding a relational system. In order to be able to process all kinds of events in the same manner, incoming events are represented by instances of the appropriate event type. They are converted into an abstract event representation. This can be thought of as an event queue scheduled in FIFO-mode. Then, the system looks for a rule to handle the event. Should the knowledge engineer have not specified a rule for the current event, the operation in question is carried out in the traditional manner. The premise of the successful termination of the transaction is not only the feasibility of the specified action but also the fulfilment of the postcondition. The action part of such a rule can contain further write-operations for the database (mostly formulated in the DML of the underlying DBMS). As a result, complex operations can be planned. In this case, it must be guaranteed that no infinite sequences occur. This can be achieved be restricting the meta model to non-recursive data structures or by logging and checking all subtransactions before a long transaction is commited. The long transaction must be rolled back if a direct or indirect cycle of operations is detected. The mechanism (shown in Figure 7 of [9]) can be simplified as soon as active databases become commonly available. The control flow can then be directly mapped onto the rule-processing components of ADBMSs. In order to describe all conditions of semantic integrity independently of a special DBMS or a certain data model, we introduced a meta model to capture complex data structures.

7.   THE META MODEL OF KNOWLEDGE BASES

The meta model consists of meta objects. Each meta object describes the structure of an object (persistent/transient). On this level of abstraction objects which are planned for retention in the knowledge base and objects which represent components of the real embedded system are treated in the same way. The basic elements are the following:

Set	Tuple
$\{\ \}: VS \rightarrow set(VS).$	$[\ ]_{a1,...,an}: VS\ 1 \times \ldots \times VS\ n \rightarrow tuple(a_1: VS_1, \ldots, a_n: VS_n).$
List	Array
$<>: VS \times list(VS) \rightarrow list(VS).$	$[]: VS^n \rightarrow array[n]\ (VS).$

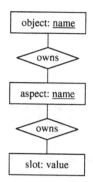

We define an algebra whose elements are the meta objects. The algebra should be minimal and extensible. The following operations are used for the construction of meta models from meta objects:

```
CreateModelObject
RenameModelObject
DeleteModelObject
InsertModelAspect
RenameModelAspect
DeleteModelAspect
InsertModelSlot
SetModelSlot
DeleteModelSlot
```

Figure 7: Elements of the meta model

Additionally, we describe the scalar types of the DBMS. The data type is one aspect of the meta object. Each aspect owns a variable number of slots containing the specific information.

Relationships between objects are modeled by meta objects whose data types are references. A reference is a type defined by enumerating a set's instances of another type currently existing in the database. In this way all conditions of semantic integrity can be formulated as a description of states in the database. The description is independent of the data model and uses the Boolean algebra and the first order logic in addition to the operations of the meta model.

Let us remember the example in Figure 3. If a component has a number of ports and the definition in a descriptive language looks like:

```
EBNF   port_def = "port" "(" port_item
                         { ";" port_item } ")" ";".,
```

then we can describe the corresponding persistent objects in the meta model as follows:

```
CreateModelObject ("port_def")
InsertModelAspect ("port_def", "Constructor")
InsertModelSlot   ("port_def", "Constructor", 1, 2)
SetModelSlot      ("port_def", "Constructor", 1, "List")
SetModelSlot      ("port_def", "Constructor", 2, "port_item")

InsertModelAspect ("port_item", "E")
InsertModelSlot   ("port_item", "E", 1, 1)
SetModelSlot      ("port_item", "E", 1, "Delete-port_item")
InsertModelAspect ("port_item", "C")
InsertModelSlot   ("port_item", "C", 1, 1)
SetModelSlot      ("port_item", "C", 1, "card(port_item) < 1")
InsertModelAspect ("port_item", "A")
InsertModelSlot   ("port_item", "A", 1, 1)
SetModelSlot      ("port_item", "A", 1,
                     "DeleteObjekt(port_def)")
```

$\Sigma_{port\_item} > 0$

The ECA-rule requires the deletion of the port definition if the number of elements in the list `port_item` becomes less than 1. On a higher level of abstraction the knowledge engineer can describe the valid states of the database.

We refer to the example in Figure 4 again. A demand may consist in the fact that structure equations only refer to the ports of classes which are in *has parts* relationship to the current class. Making use of meta objects regarding the class, the components of the class, and of the ports, we formulate the state condition.

```
∀ a ∈ database_def.class_def,
 ∀ b ∈ a.property_def.structure_def,
  ∀ c ∈ b.struc_item.equation.str_port_def.str_single_port,
   ∀ d ∈ database_def.class_def,
    ∀ e ∈ d.property_def.port_ref,
     ∀ f ∈ a.property_def.part_def:
      ∃ c.part_identifier = f.task_name ∧
        c.str_task        = ⊥              ∧
        c.port_ref        = e.port_name ∧
        d.str_port        = ⊥              ∧
        f.subclass        = d.class_name
```

The following events can be derived from this representation:

```
On-Update Class, On-Update Task, On-Update Port
On-Delete Class, On-Delete Task, On-Delete Port
```

If one of these events occurs, the condition must be tested again. In the condition section of the ECA-rule the fulfilment of the terms outlined above is tested. The action section depends on the kind of the initiating operation. An *update*-operation will not start if its results hurt the state condition.

A *delete*-operation (e.g. for `Port_A` or `step2`) can necessitate further interventions to restore the consistency of the data base. Consequently, the whole structure equation:

structure (step2.Port_A := step1.Port_C);

must be removed.

8.   USAGE OF THE META MODEL

The meta objects can be represented in a flat file, a database, or in the XML format. Two essential categories can be derived from the meta model: the schema definition of a concrete database and the lot of the ECA-rules. Because each DBMS has its own mechanisms for the preservation of the logic integrity, the underlying principles must also be described in the form of meta objects.

In the case of relational databases the schema definition consists of CREATE TABLE-Statements (SQL) including constraints which are generated based on logical conditions and are formulated in the meta model. In the case of object-oriented databases the special DDL of the target system is used. An example is given in [9].

9.   CONCLUSION AND OUTLOOK

In this paper we presented a new approach regarding the construction and maintenance of knowledge bases for embedded systems, which is based on an extensible meta model. As a case study, a knowledge base for the domain of the robot controllers was developed [12].

The customer is able to build his own database for the administration of draft data. Since rules affecting the database can be defined independently of the applications, any commercial tool can be included in the process of development. After the procedures of active rules have been proven to be working on relational databases, the implementation on a concrete object database starts. We plan to extend this approach so that hardware and software can be simulated under event control in the same manner.

References

[1]   J. Böttger, K. Agsteiner, D. Monjau, and S. Schulze,
        An Object-oriented Model for Specification, Prototyping, Implementation and Reuse.
        In: *Proceedings of Design Automation and Test in Europe Conference DATE 98*,
        Paris (France), Febr. 23-26, 1998, pp. 303-310.

[2]   Bryon K. Ehlmann,
        A Data modeling tool where associations come alive.
        In: *Proceedings of the IASTED International Conference Modelling, Identification and
        Control (MIC 2002)*; pp 66-72; February 18-21, 2002, Innsbruck, Austria,
        ISBN: 0-88986-319-9, ISSN: 1025-8973.

[3]   T. Kuhn, T.Oppold, M. Winterholer, W. Rosenstiel, M. Edwards, and Y. Kashai,
        A Framework for Object Oriented Hardware Specification, Verification, and Synthesis.
        38th Design Automation Conference (DAC), Las Vegas, USA, 2001, pp 413-418,
        ISBN 1-58113-297-2

[4]   D. Monjau, M. Sporer,
        Knowledge-based design of embedded systems.
        In: *International Journal of Knowledge-Based Intelligent Engineering Systems
        Engineering Research Centre, School of Engineering, University of Brighton*
        Pages 1-8; January 2002, Vol. 6, No. 1, ISSN: 1327-2314.

[5]   R. Seepold, N. Martinez Madrid (Eds.); Virtual Components Design and Reuse
        Kluwer Academic Publishers, 2000, ISBN 0-7923-7261-1.

[6]   M. Sporer, K. Agsteiner, D. Monjau, and M. Schwaar,
        Knowledge Based Specification and Modeling of Embedded Systems.
        EUROMICRO 99, Workshop on Digital System Design. Milan (Italy),
        September 8-10, 1999, IEEE Computer Society, ISBN 0-7695-0321-7, pp. 398-401.

[7]   D. Monjau, M. Sporer
        Reuse-Oriented Design of Embedded Systems.
        In: *Knowledge-Based Intelligent Engineering Systems & Allied Technologies
        KES'2000*, University of Brighton, Sussex, Volume 2, page 691-694,
        Brighton, U.K., 30.8. - 1.9.2000, ISBN 0-7803-6400-7.

[8]   M. Varea, and B. Al-Hashimi,
        Dual Transitions Petri Net based Modelling Technique for Embedded Systems
        Specification. In *Proceedings of Design, Automation and Test in Europe*, Munich,
        Germany, 13-16 March 2001, pp. 566-571.

[9]   D. Monjau, M. Sporer
        Ensuring semantic integrity in knowledge bases for embedded systems
        In: Proceedings of Forum on specification & Design Languages, Vol. 2;
        September 24-27, 2002 - Marseille, France

[10]   http://www.vsi.org/library/vsi-or.pdf

[11]   http://herkules.informatik.tu-chemnitz.de/wisyra/CLINT_syntax.html

[12]   http://herkules.informatik.tu-chemnitz.de/wisyra/rodos.html